D1050417

Technical Writing Basics

Technical Writing Basics

A Guide to Style and Form

Third Edition

Brian R. Holloway
Mountain State University

PEARSON

Prentice
Hall

Upper Saddle River, New Jersey
Columbus, Ohio

Library of Congress Cataloging-in-Publication Data

Holloway, Brian R.
  Technical writing basics : a guide to style and form / Brian R. Holloway. -- 3rd ed.
    p. cm.
  Includes index.
  ISBN 0-13-114089-2
  1. English language--Technical English--Handbooks, manuals, etc. 2. Technical writing--Handbooks, manuals, etc.   I. Title.

  PE1475.H65 2005
  808'.0666--dc22

2004005609

Editor in Chief: Stephen Helba
Senior Acquisitions Editor: Gary Bauer
Editorial Assistant: Natasha Holden
Production Editor: Louise N. Sette
Production Supervision: *The GTS Companies*/York, PA Campus
Design Coordinator: Diane Ernsberger
Cover Designer: Jim Hunter
Cover art: Photo Disk
Production Manager: Pat Tonneman
Marketing Manager: Tim Peyton

This book was set in Gill Sans by *The GTS Companies*/York, PA Campus. It was printed and bound by R. R. Donnelley and Sons Company. The cover was printed by Phoenix Color Corp.

Pearson Education Ltd.
Pearson Education Singapore Pte. Ltd.
Pearson Education Canada, Ltd.
Pearson Education—Japan

Pearson Education Australia Pty. Limited
Pearson Education North Asia Ltd.
Pearson Educación de Mexico, S.A. de C.V.
Pearson Education Malaysia Pte. Ltd.

10 9 8 7 6 5 4 3 2 1
ISBN 0-13-114089-2

Foreword

The ability to write means empowerment. It means that you can present your ideas within a format to which others will respond. This book is for anyone learning about writing in the workplace and the materials used in seeking employment. The models of style and format will help you whether you are taking this subject as a course or whether you study on your own. You do not have to be a student in a technical writing class to use this guidebook.

Reading this text, you will learn about the special writing featured in business, agencies, or other public settings—writing that must produce concrete results and that requires definite protocols for the proper exchange of information.

The content and approach of the second edition also characterize this volume. Of course, I have altered material to keep the book current, to answer needs, and to respond to suggestions made by friends in education, technology, and management. I hope that this new book will assist you, and I welcome you to the third edition of *Technical Writing Basics*.

—Brian R. Holloway

To the Student . . . and Teacher

To be an effective writer in the workplace, one must know the expectations. Special characteristics govern writing within, to, or for businesses, social-service agencies, health-care providers, and government entities. Often, textbooks surveying the challenges of such writing have stressed one of its several functions. For example, some authors teach entirely from a "technical" perspective, limiting discussion to the mechanics of production. Manuals depicting the proper way to construct tables, pie graphs, and charts are valuable resources; however, they supplement—rather than explain—the process of communication in the workplace. Similarly, a text focusing entirely upon the presentation of data may create an impression that one does not use persuasive techniques when informing an audience. Yet most students know that even the driest report may be made appetizing if its "package"—the format—is enhanced. Communication theorists, in fact, have a difficult time determining where "information" ceases and where "persuasion" begins; the two overlap, rather than comprise the ends of a continuum. A third editorial choice made by writers of texts focuses upon the persuasive aspects of communication in the workplace, as if the modes of presentation mattered most.

This text draws from each of these partial perspectives in surveying holistic challenges within business and technical writing. Because this book is intended for students who have taken a freshman composition course but who have not necessarily worked in fields that demand the use of business and technical writing, its examples—real or fictionalized—are practical and basic. Writing letters of application and adjustment, constructing informational and persuasive reports, and encapsulating material so that it can be convincingly communicated are all activities shared by most college students; therefore, this book derives much of its illustration from such models. Throughout the text, then, three goals drive the content:

1. Students should study the requirements of informative writing.
2. Students should learn how the techniques of persuasion operate in writing in the workplace.
3. Students should practice casting informative and persuasive writing into an appropriate format.

Because what we practice not only depends on different aspects of business and technical writing but transcends it, I recommend that we call such communication *transactional writing*. In a transaction, the communicator provides information to the recipient of communication, but the recipient often must give up something as well: a prejudice (against the action proposed), free time (which could be spent eating lunch instead of reading a memo), or a method of doing something (which the information just received happens to contradict). Frequently, one must offer the flattest data in terms calculated to make reading a report seem worthwhile. Getting the other person to read one's material, presenting such material clearly and accurately, and adhering to standards of format acceptable in the field become the goals of transactional writing.

Educational Approach

The exercises and assignments in this text build in complexity, chapter by chapter, as well as inside such chapters; for example, in Chapter 7 the preliminary report (Figure 7.14) contains material that can be used in the proposal (Figure 7.15), and the final report (Figure 7.16) includes imports from the proposal itself. Such an incremental, cumulative approach assists students who use computers in their writing, as saved material that constitutes a previous assignment can be retrieved, modified, and transferred to the new document. Should the final report be collaborative, students can integrate their reports on disk as well. Such a collaborative method might encourage students in similar fields to work together to create unified projects. This approach reaffirms the future value of the work the student has just completed, and is pragmatic—real examples and models demonstrate what should be done.

Synopsis of the Table of Contents

This text is conceptually divided into three sections. The first unit, the "apprenticeship," covers the basics of transactional writing and encompasses Chapters 1 through 3. Chapters 4, 5, and 6 expand the scope of technical and business writing, building on the foundation established earlier but introducing more complex assignments. Chapters 7 and 8 put the acquired knowledge to work in creating two polished, multisectioned documents: the formal report and the job portfolio. A short list of references and an appendix on graphics follow.

Chapter 1—Introduction

The first section in Chapter 1, *What Is "Business" or "Technical" Writing?*, surveys the features of transactional writing, exploring differences between such writing and the academic prose taught in composition classes. It discusses working together in small groups to achieve a writing goal, growing a larger document from a smaller one—or from fragments—and using computers to assist the process of creating a document.

Saying What You Mean surveys the basics: grammar and usage. It identifies sentence types and problems, focusing on the impact of phrasing but discussing other mechanical issues as well. A section on diction encourages writing within the context of "world English," avoiding localisms and expressions that might confuse or antagonize readers.

Chapter 2—Organizing Information

Chapter 2 discusses how we impose patterns of logical order upon the infinite field of data around us, selecting from this field that which is necessary to support our message. This chapter reviews specific templates that structure logical presentations—many of them called "modes" in writing texts—such as *summary, process, analysis, comparison,* and *persuasion.*

Chapter 3—Letters, Memos, E-mail, and Related Forms

Chapter 3 presents a simple organizing framework of business communication: the message-support-closure framework. It next discusses the features of letters, memos, and transmittal documents, drawing on the writing patterns analyzed in Chapter 2.

Chapter 4—Our House to Yours: Using Summaries to Inform

Chapter 4 is a respite from all that memo and e-mail writing, but surveys documents similar in form: bulletins, descriptive leaflets and flyers, and public service announcements. This discussion expands the work with *summary* begun in Chapter 2.

Chapter 5—Directions and Instructions: Writing About Process

Chapter 5 focuses on *process* writing: informative and persuasive documents explaining how to do something or how something gets accomplished. The chapter surveys posted directions, instructional pamphlets, and other examples.

Chapter 6—Using Analysis: Writing a Report

Chapter 6 explores *analysis* and its ally, *comparison.* It suggests ways to use the templates introduced in Chapter 2 and develops techniques of ensuring continuity when writing analytically. This section offers tips on constructing projects in groups.

Chapter 7—Writing the Formal Report

Chapter 7 puts into practice all the skills learned while using this book; the templates studied reappear as parts within a larger concept and assume subordinate roles within that bigger structure. This chapter also covers research techniques, the use of traditional and electronic sources, and writing practices. Reporting on work-in-progress leads to writing the formal proposal; this document can be expanded to construct the final report.

Chapter 8—Selling Yourself

Chapter 8 extends the art of informing and persuading to one's search for employment. This chapter covers the research involved in job seeking (both in keeping a work-history file and in tracking down leads in the library). It reminds the reader that modern job searches can be greatly facilitated by new CD-ROM tools and the Internet. This chapter then looks at the components of a job seeker's arsenal: résumés, cover letters, vitae, and portfolios. The chapter discusses interviews and follow-up correspondence as well.

References

This section lists other sources of information about business and technical writing, which will enhance and augment the work begun in this text.

Appendix—Enhancing Your Document with Graphics

The appendix surveys integrating pictorial material into your document.

Acknowledgments

This little text's long genesis owes much to the students I've taught for more than twenty years who have helped me understand the challenges of teaching writing. Their energy and insights have enhanced each class. I want to commend Mountain State University, too, for its appreciation of my project and its support of my endeavors.

Neil Manning deserves thanks for sharing his perspectives from industrial research. (I now get to repay Neil for putting me in the dedication of his engineering dissertation).

The camaraderie and hospitality of Doug, Mary Leigh, Jessica, and Jeff Burns have been outstanding. Thanks, too, Doug, for helping me learn about technical communication in the biosciences.

I'm grateful to Steve Helba, editor in chief; Gary Bauer, senior acquisitions editor; and the staff at Prentice Hall for believing in this project and for superlative guidance and support. Christie Catalano deserves thanks for encouraging me in the very beginning when I had a rough-hewn prospectus in hand.

I also thank the following reviewers for their helpful suggestions: Maryanne Brandenburg, Ph.D., Indiana University of Pennsylvania; Susan Harper, Penn State University; and Rachel Jorden, Hudson Valley Community College.

Additionally, the comments of Doug Burns, Rachel Lanier, Lee Cook, Bonniejean Alexander, Staci Craft, Vincent Massey, and Paula Fields were particularly helpful in revising this edition.

Finally, my wife Kathy, my daughter Rachel, and my mother deserve, as always, my thanks and my love.

To these people, and to all my future readers, I dedicate this book.

Contents

CHAPTER

1

Introduction

What Is "Business" or "Technical" Writing?

Business or technical writing is engineered to display information effectively on the page in order to get results, or in other words, to inform and to persuade. Such "real-world" composition differs in many ways from the academic essays assigned in college or high school English courses, though there are some overlaps. Everything written within a business and technical format is produced with the intent of achieving a targeted response, such as

◆ the understanding of information
◆ the acceptance of a proposal
◆ the consideration of a feasibility study
◆ the return of a defective item
◆ the establishment of goodwill
◆ the sale of merchandise
◆ the hiring of the writer!

Transactional Writing

This kind of communication might best be described as "transactional" rather than "business" or "technical." In a transaction, there are two parties. One proposes something to the other, but by accepting the ideas in that document the other party may have to give up something—money, time, even beliefs and values. The other party may have to devote valuable minutes—perhaps a lunch break—just to *reading* the proposal. As we'll see, the strategy of persuasion underpins this form of communication, since it is always difficult to convince people to act.

The first principle of persuasion is to know your audience. What are its needs, values, goals? No busy manager is going to have the time, for example, to wade through a leisurely essay that dawdles across four pages of dense type unrelieved by white space. Busy people—the targets of most transactional writing—by definition do NOT have the patience to decode the subtext of an intricate communiqué. This puts the burden on the writer, who must direct the reader, using headings to show the outline of the discussion and lists to clarify the points discussed (Figure 1.1). Such an approach encourages the reader to follow the document as it proceeds. A good transactional writer won't alienate an audience along the way, either, through a misguided choice of words.

The second principle of persuasion is to know exactly what you want your audience to do. Consider what would happen if you wrote a cover letter in response to a want ad—but never demonstrated in the letter that you'd like to be hired for the job! Personnel officers in large and small corporations see hundreds of such letters.

Figure 1.1   One Example of a Technical Format
Using Headings and Lists to Direct the Reader

To:	**Recipient**
From:	**Sender**
Date:	**March 23, 2004**
Re:	**Subject**

Bold Heading
Text of first paragraph, beginning flush left—no indentation. xxxxxxxxxxxxx
xxxxxxxxxxxxxxxxxxxxxxxxxxxxxxxxxxxxxxxxxxxxxxxxxxxxxxxxxxxxxxxx
xxxxxxxxxxxxxxxxxxxxxxxxxxxxxxxxxxxxxxxxxxxxxxxxxxxxxxxxxxxxx

Bold Heading
Text of second paragraph. xxxxxxxxxxxxxxxxxxxxxxxxxxxxxxxxxxxxxxxxxx
xxxxxxxxxxxxxxxxxxxxxxxxxxxxxxxxxxxxxxxxxxxxxxxxxxxxxxxxxxxxxxxxx
xxxxxxxxxxxxxxxxxxxxxxxxxxxxxxxxxxxxxxxxxxxxxxxxxxxxxxxxxxxxxx

Bold Heading
Text of third paragraph using bulleted lists:

• First point
• Second point
• Third point

(Subordinate points expressed as bulleted lists provide visual relief and guide the reader, as well).

Bold Heading
Text of conclusion. xxxxxxxxxxxxxxxxxxxxxxxxxxxxxxxxxxxxxxxxxxxxxxxx
xxxxxxxxxxxxxxxxxxxxxxxxxxxxxxxxxxxxxxxxxxxxxxxxxxxxxxxxxxxxxxxxx

Third, you must use clear and specific content within a business and technical format. Doing your research and convincing the reader with detail is only half the challenge—you must also adhere to the page layout and style of presentation expected in order to get results. We'll discuss patterns and formats later.

Transactional Versus Academic Writing

Since you're probably entering this course after having taken English Composition, you'll no doubt note many things you learned in that class that transfer to business and technical writing. Both types of communication require clarity, focus, audience-awareness, development, coherence, and smooth expression—and the absence of problems with word choice and grammar.

But there are many differences, which you'll notice as we work through the text. A few striking ones include

◆ An "outline" form of presentation instead of an essay form
◆ The use of different fonts and sizes of type to create eye relief
◆ The frequent use of flush-left text instead of tabbed indentations at the beginning of paragraphs
◆ Single-spaced paragraphs separated by double-spacing
◆ Different styles of format

Working Together

One of the biggest challenges in business and technical writing is that frequently the different sections of a document are contributed by different people, each having particular stylistic quirks, and each emphasizing some things that may not be important to the document as a whole. Then the group must decide how to reconcile all the parts with each other, what to enhance, what to discard, and how to integrate parts so that they become a seamless unity.

Growing a Document

In group situations, documents are often "grown" from small, isolated sections into developed, multiunit presentations: major proposals, employee handbooks, feasibility reports. Much give-and-take and many hours of reexamining the drafts produce a finished text. Many writers begin with an outline that shows the major headings in place; the supporting material under each heading is developed separately by different people; then the whole package is put together, reviewed, and re-reviewed. What looks like a natural, effortlessly produced presentation is really the product of intensive work (Figure 1.2).

Figure 1.2   Skills Required to Develop Documents Within Groups

Teamwork. Discussion must move beyond debate and disagreement to a consensus produced by mutual agreement. No one should feel "left out" of the decision making, nor should a few seek to dominate and "overpower" others.

Diversity. All approaches to the problem must be considered. Frequently there are several ways to organize the same material, and several topics under which the material may be organized.

Project Management. The group must set reasonable deadlines to complete its tasks. It must identify the skills of its members and put the appropriate members in charge of the sections of the project in which their skills will best be used. It often happens, for example, that one or two people become the final editors of a document.

Invention/Development Strategies. The group must use
- *Associational thinking* to explore all possible topics and approaches
- *Focused thinking* to unify selected topics under one central concept, creating transitions to link topics together
- *Reflective thinking* to assess the effectiveness of the unified presentation, and to make appropriate changes when necessary

Using Computers

Writing by yourself, you'll undergo the same challenges. It's easiest to develop a document on disk, using a computer program with which you are familiar. Big projects often begin with material in separate files, which can, once developed, be imported into the main shell of the document. Figure 1.3 illustrates the pattern of document development.

There are other good reasons to use computers in drafting:

◆ *They make proofreading easier,* since the text looks clear and programs can check your spelling.
◆ *They facilitate recursive writing*—that is, they let you return repeatedly to the document without having to begin anew.
◆ *They allow you to prepare alternate versions* of the same project so that you can pick the best one.
◆ They are unsurpassed in providing options for attractive format.

Current studies question whether computers can actually help your thinking as you compose, but this, too, may be a benefit for you.

Figure 1.3 A Pattern of Document Development

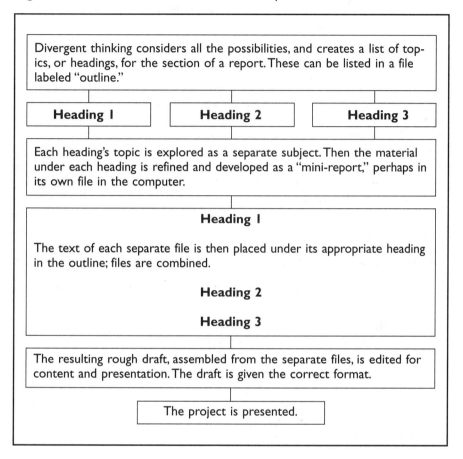

Divergent thinking considers all the possibilities, and creates a list of topics, or headings, for the section of a report. These can be listed in a file labeled "outline."

| **Heading 1** | **Heading 2** | **Heading 3** |

Each heading's topic is explored as a separate subject. Then the material under each heading is refined and developed as a "mini-report," perhaps in its own file in the computer.

Heading 1

The text of each separate file is then placed under its appropriate heading in the outline; files are combined.

Heading 2

Heading 3

The resulting rough draft, assembled from the separate files, is edited for content and presentation. The draft is given the correct format.

The project is presented.

Saying What You Mean: A Grammar Refresher

Evidence suggests that when a single piece of writing is produced by several people, problems of style and unity not apparent in the work of separate authors may surface. The authors must ensure that their document speaks with a single voice and in a manner free from distractions, especially at points in the text where the different contributions are joined. In fact, a large document created by a single writer may contain similar disruptions, because parts of the text created earlier might have a different style than those produced later. In addition, presenting more complex ideas puts greater pressure on the grammar that conveys them. Thus, though grammar and expression may not ordinarily be problems for you, the following review may provide a helpful reminder of basic principles upon which to rely.

Grammar and Usage Assist Credibility

A misplaced comma or a misspelled word might cost your report credibility, and the wrong word choice, or usage, might baffle or alienate your reader. Result-oriented writing cannot afford such mistakes. Your standard writing equipment should include a good dictionary (such as *Webster's* or *American Heritage*), a current manual of style and grammar such as the *Prentice Hall Reference Guide to Grammar and Usage* (by Muriel Harris), and a thesaurus, or dictionary of synonyms. Programs such as "Spell-Check" or "Grammar-Check," though helpful, will not save you; you most certainly don't want a piece of software thinking for you in critical matters of communication. Too many silly sentences are produced by writers with too much faith in machines. Figure 1.4 illustrates the limitations of spell-check programs.

Now consider the following example:

> Too over come there problem, a person must have faith in the therapist treatment.

Here in this small sentence lurk many errors software may not catch. A spelling program would see no misspelled words, and a grammar program would not know what to do to make the words make sense. "Too" must become "To," "there" must become "their"—or is this supposed to be a pronoun referring to "a person"? In that case, "your" might be the proper word to choose. "A person"—normally an awkward phrase—might then be replaced with "you." And the phrase "therapist treatment" reads as if the therapist is

Figure 1.4 Spell-Check Will Not Save You!

Words Typed	Suggested Word
fractal	factual
workforce	workhorse
downsizing	downswing
sketcher	sketchier
storyboard	scoreboard
refocusing	refusing
HVAC mechanicals	HAVOC mechanical
admin. ass't.	admin. asset
TQM	TAM
the Demings Concept	the Denims Concept
raster	rasher

getting treated! What we really want is to use the appropriate words and phrases, so that the sentence reads like this:

> To overcome your problem, you must have faith in the treatment prescribed by the therapist.

The non-grammar and non-sense you've just seen is actually typical of confusing technical writing. You may have tried to follow the assembly instructions that came with a bicycle, a gas grill, or an ice-making attachment for a refrigerator, only to discover that the grammar of the directions made them incomprehensible. At the least, credibility is destroyed, as you can see in the example illustrated in Figure 1.5. Here, despite good intentions, something has gone awry.

Exercise

Figure 1.5 is a simulation of the cautionary notice inserted in packages of blank videotape—an insert that can use some editing. Can you supply some editorial guidance?

Figure 1.5   Strange Process Instructions

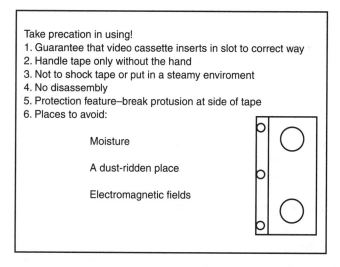

Take precaution in using!
1. Guarantee that video cassette inserts in slot to correct way
2. Handle tape only without the hand
3. Not to shock tape or put in a steamy enviroment
4. No disassembly
5. Protection feature—break protusion at side of tape
6. Places to avoid:

    Moisture

    A dust-ridden place

    Electromagnetic fields

Note well:

This document appears to be written in great haste and printed on the back of a carton that no one ever reads. But the uninitiated will have to read an instructional document like this to learn how to use a videocassette. As an editor and writer, your job is to take this rough material and turn it into a clear set of guidelines.

How was your editing experience? You were right in noticing oddities of spelling, repetition, and word selection here. What is a "precation"? How else can one handle something but with a hand? How is moisture a "place"? Now rewrite this document so that it communicates information clearly, but still uses words sparingly. Next, try your editing ability on the following extracts from documents.

Exercise

Can you rewrite these excerpts from technical reports so that they communicate clearly? Use your handbook of grammar and style to help you; consult your dictionary or thesaurus as needed.

1. The question, stated as such it is in the social services realm, is perhaps not the true idea which I should expound upon.
2. This short list hardly at all includes all the reasons for obtaining a skill in writing if one is in a health field.
3. The assistance program. To give a man a choice, whether to father a family with a working wife, a career spouse, or not marrying at all.
4. Much time is wasted when a person trying to explain or illustrate a point is unable to do so without using alot of worthless words or phrase.
5. Poorly placed commas and unprecise words will make a technical writing dull and inaccurate, and I could leave the reader puzzled instead of enlightened.
6. Paragraphs: Both types of use the same tools, grammar, organization.
7. The planners reviewed cities back to back all over the country.
8. The materials on the Internet are like bricks: they all build together to join a fountain of information.
9. Of course there have been many brilliant breakthroughs throughout history like Einstein and Galileo.
10. S/he may find the doctor/s not interested in her/him but only concerned with his/her disease/s when one comes/goes there.

Your grammar handbook will provide particular advice about the writing problems above, and there are many possible solutions for each tangled sentence. Though some selections were the unfortunate results of pained deliberation, most of the previous examples were written in haste, carelessly, and suffered for it. Be sure to proofread your own work as thoroughly as you've dissected these sentences, using the guidelines given in Figure 1.6. Many writers are more adept in proofreading the work of others than they are skilled at assessing their own productions.

Figure 1.6 Proofreading Advice

Read for content. First, make sure that the document is supported and that its mission is clear. All documents should have unity (a clear goal), development (convincing illustration and logic), and coherence (transitions and phrasing must be appropriate); THEN—

Forget the content. Your "residual memory" of the draft will evaporate if you put it aside (for a day or two if possible). Then you can look at the document with fresh eyes.

Isolate each word. Cut a ⅛-inch × 1¼-inch slot in a 3 × 5 index card. Place the slot over each word in your text. Without distracting context, misspellings and other problems may become visible.

Disrupt perceived continuity. Proofread backwards, paragraph by paragraph, or word by word, using the proofreading card if necessary.

Sentence Types in Technical Writing

If you've forgotten the intricacies of what Mr. or Mrs. Grundy taught you in eighth-grade grammar, use this simple guide to sentence construction in order to improve your own transactional writing (those skilled in sentence construction may still want to review the following material):

◆ *A sentence consists of a complete thought expressed as a subject and verb; other words may be added to make the thought specific.* Examples:

I slept.
I slept late.
I slept late because the idiot next door played death metal music until four a.m.
I slept late, missed the exam, and I'm proud of it.

◆ *All sentences are built of a few basic components, called independent clauses, dependent clauses, and phrases.*
◆ *A simple sentence—subject plus verb—is an independent clause:*

I ran.
I ran over the field.
I ran over the field to Grandmother's house.
I ran over the field to Grandmother's house to borrow money for a down payment on a new Camaro.

Notice that an independent clause (a subject plus a verb) may have descriptive *phrases* attached to it. "Over the field to Grandmother's house to borrow money for a down payment on a new Camaro" is a collection of such phrases.

◆ *A dependent clause can't stand by itself but must attach to an independent clause to provide meaning.* Such clauses are frequently noticeable because *transition words,* or words that link thoughts, introduce them. Here are examples of dependent clauses:

> Although it rained today
> Because the growth factors aren't measurable
> However the formula turns out

◆ *All sentences other than simple ones are built by selecting independent and dependent clauses to combine in different ways.* A *simple sentence* uses one independent clause. A *complex sentence* combines an independent clause with a dependent one. A *compound sentence* combines two independent clauses using a conjunction—"and" or "but" for example—or a semicolon. A *compound-complex sentence* has a dependent clause and two independent ones. Here are examples of the four sentence types:

> I ordered the pizza. [Simple]
> Though I was dieting, I ordered the pizza. [Complex]
> I ordered the pizza and I bought a bottle of pop. [Compound]
> Though I was dieting, I ordered the pizza and I bought a bottle of pop. [Compound-Complex]

Get to know these sentence patterns; you need the different types to express thoughts in technical writing. For example, a simple sentence might state a concept that needs to be read rapidly. Complex sentences will relate two unequal thoughts and frequently act as transitions in documents. And compound sentences coordinate two equal thoughts, giving equivalent impact to both. You can avoid repetition in transactional writing by using compound and complex sentences.

Sentence Problems

As you remember from English Composition, failure to recognize the different sentence forms and their building blocks is what causes most grammatical errors. A dependent clause masquerading as a sentence is a *fragment.* Two independent clauses run together without any punctuation constitute a *fused sentence.* Two independent clauses pretending to be glued together with a comma make a *comma splice.* Stringing a bunch of independent clauses together with ands and buts produces the infamous *run-on sentence.* Take a look at these examples:

> Although the rain won't fall. [Fragment]
> The rain won't fall the sun won't shine. [Fused]
> The job was mine, I volunteered for it. [Comma Splice]
> The job was mine and I volunteered for it and no one will take it away from me. [Run-on]

Try out your understanding of sentence structure in this next assignment.

Exercise

Revise these ten selections so they make sense as coherent sentences. There are many possible solutions!

1. Pup tents are designed for sleeping and allow little room but feel cozy.
2. A large umbrella tent is more expensive. Although roomier.
3. Mom bought the car in 2003 it was driven only 19,000 miles.
4. Once you settle on what college. You can then explore the possibility of a Stafford loan.
5. These calculators provide the same features as others, they also plot points on a graph.
6. Inspect the axles, there should be no bends.
7. Release the brake levers they should not bind.
8. The inside of the case should be shaped to hold the guitar snug and should have durable latches and comfortable handle placement with a covering of waterproof Tolex.
9. Bamboo tent poles may crack. Although lighter in weight.
10. The customer refused the accessory package, it did not include pinstriping.

The purpose of this review of sentence construction is to remind you of the basics. The units of meaning in technical writing are well-constructed sentences, and from these units you can build anything.

Other Mechanical Difficulties

There are two other grammatical problems that can persist in writing and that can damage a transactional communication. These are dangling modifiers and nonparallel construction.

Dangling Modifiers: A *dangling modifier* occurs when the phrase or clause that needs to modify a part of a sentence is not near that part. Sometimes, the phrase or clause modifies something *outside* the sentence entirely! You can hear such confusing expressions on the nightly televised news, or study and correct the following examples.

1. They emptied the cans, who are my brothers in the garbage business.
2. Walking through the store, French bread loaves were two for one.
3. Boiled, baked, or roasted, you can't outdo the flavor of chicken.
4. When eight years old, Granddad bought me a toy train.
5. He published his essay on primary campaigns in Missouri.
6. The fugitives used a site by a cave not frequented for years.
7. They were warned to leave by the ranger strongly.
8. She looked at the woman with a problematic expression.
9. With instructions, insert the applicator into the ear.
10. Insert the applicator into the ear with the instructions.

Parallelism: *Parallelism* merely means that all elements in a list or series need to be of the same grammatical kind. For example, if the first two phrases are nouns, the third phrase must also be a noun. If the first word in a descriptive list is an adjective, the next two must be adjectives also. Breaking this pattern confuses the reader. Here is an example of parallelism at work:

"He was a greedy, conniving, shifty politician."

Note that the words describing "politician" are each adjectives. Because business writing uses many lists—and because these lists follow parallel structure—understanding this principle is important. You want this—

New hires should

◆ Report for work
◆ Sign the log
◆ Obtain the tax forms

You do not want this—
The popular conception of used-car salespersons is that

◆ Insincere
◆ They will rook you
◆ Plaid suits are seen on some of them

Figure 1.7 illustrates an example of parallelism.

Figure 1.7   An Example of Parallelism

In the article below, notice that the items in the bulleted list each begin with an active verb—enact, create, establish, and train.

Reducing Government Secrecy

A federal commission has recommended that the United States:

• Enact a federal law that would set detailed procedures for classifying documents that affect national security and for declassifying records that do not. Would declassify most records 10 years after their creation, and release all but the most sensitive after 30 years.
• Create a National Declassification Center at an existing federal agency, such as the National Archives and Records Administration, to oversee the release of government records.
• Establish a single executive-branch office on declassification that would replace existing offices and report to the National Security Council.
• Train federal employees to classify records more selectively.
• Establish ombudsman offices within agencies to resolve classification issues.

From Martin, Stephen. "Federal Panel Seeks to End 'Culture of Secrecy' That Restricts Historical Records." *The Chronicle of Higher Education* 14 Mar. 1997: A28

Exercise

Correct the parallelism in the following statements.

1. Turn on the lights, sweep up the shredded paper, and you should not touch the "Initiate Nuclear War" button.
2. The species may be found under rotting logs, beneath leaves, and some people see them when they lift up small, smooth stones which are usually round.
3. I can supply the talent you seek in the following areas: microwave transmission, I can maintain thyristors, beam-focusing.

Diction and International Concerns

Your choice of words is as important as adhering to the rules of grammar. A bossy tone turns away loyal readers; the irrelevant use of specialized terms irritates nonspecialists; clichés prevent understanding by filling up space with meaningless phrases; inappropriate word choice (including unconscious punning) deflects attention from the issue presented and may offend.

You can recognize authoritarian language in such phrases as "It has come to my attention that . . ." or "Zero tolerance of deviation from this procedure will be implemented." Note that this first example focuses on the *sender* of the message rather than the recipient, and that the second example uses an impersonal command in passive voice—who is implementing the procedure, and why is he or she hiding behind this language? Avoid unnecessary use of the *passive voice,* in which the subject of the clause becomes the object and the object the subject. Not "The ball was thrown by Tom" but "Tom threw the ball." Not "The failed program was implemented" but "We implemented the failed program." Passive voice does tend to sound pompous or unnecessarily obscure, as it removes the "doer" of the action from its normal word order.

Misapplied technical language, or jargon, occurs when a writer uses overly specialized language to communicate with a general readership or when the language of one field is used inappropriately in another. For example, "The appropriate remuneration is of course desired" might sound passable in a law office, but is not a good way to ask your boss for a raise. And you'd be unlikely to give directions to your house by telling people to look for "lot 25, surface rights only, situated at the NW corner of a rock declivity abutting survey line running E along Rock Ridge Road."

Some jargon words have become so overused that they have lost any clear meaning. A "venue," originally the locale of a trial, now is used to mean any location at all. Thus a hairdresser, a rock band, and a film might all have "venues." And what do phrases such as the following mean?

maximal environmental friendliness

differing functional orientation

viable programmatic implementation

growth-driven objectives

There do not seem to be clear equivalents for these statements; rather, they seem designed to be plugged into bureaucratic documents at random. If you can't picture the meaning of a phrase clearly in your mind, change the phrase until you can.

Clichés are commonly used fillers in conversation. Like jargon-choked language, cliché-ridden text lacks meaning. *For example, although you may want to throw caution to the winds, and even if you are the salt of the earth, your tendency to add insult to injury must be nipped in the bud or you will burn your bridges and leap out of the frying pan into the fire. Then the fickle finger of fate will force you to lay it on the line and put all your cards on the table, in which case you may have to look over your shoulder for the long arm of the law as you bite off more than you can chew.* Convinced? Don't rely on clichés to lend a "folksy" tone to your correspondence—you'll discover that readers find it phony. Again, visualize what you want to say, then say it precisely.

A striking instance of *inappropriate language* is profanity. Other manifestations of inappropriate language include subconscious wordplay, such as inadvertently calling a woodworking apprentice a chip off the old block, or asserting that a sewer tax won't drain one's finances. Such punning is often associated with clichés.

Sexist language is a species of inappropriate language: a memo referring to executives as exclusively of one gender is sure to irritate. The English language contains options for dual-gender and gender-neutral writing. These include

♦ Using plural forms to include both genders: "All must keep phone logs in this business."
♦ Using the neutral "one" in place of gender-specific pronouns—not "Everyone has his phone log," or the questionable "Everyone has their phone log," but "In business, one keeps a phone log."
♦ Rewriting the text to eliminate gender—"Every manager of this company should keep a phone log." This strategy eliminates the rough reading that another choice, the split pronoun, frequently causes: "Those are the options available to him/her" is considerably more awkward than "Those are the options available."

No racial, ethnic, or sexual derogation should ever appear in your writing.

Your writing must also take *international concerns* into account. English is a world language—more than five hundred million users and growing—so removing local expressions from international correspondence becomes critical. Such local expressions as clichés, references to American athletics, and certain illogical phrases called idioms can make no sense to readers for whom English is not a native language. "World" or "international" English—a trade language and the discourse of computer information—must remain free of localized encumbrances to work. For example:

♦ *Clichés.* "Throwing the baby out with the bath water" may not only lack meaning—it may sound downright gruesome!
♦ *Sports.* "He'll be the mightiest Sultan of Swat since the Babe" not only seems meaningless, but contains an implicit ethnic stereotype.

◆ *Idioms.* Every language contains expressions that are not necessarily logical by grammatical standards, but which have evolved to convey meaning nevertheless. "How are you?" is used as a greeting by people who have no intention of receiving a logical response—it's just a tag phrase to lubricate conversation. "Parking lot" is jarring to British English users accustomed to "car park"—and vice versa. The American who boasts of getting a good deal on an appliance by saying "I stole it" might convey a different meaning to international users. And what does the expression "getting a good deal" literally mean? Anything that sends your readers to their dictionaries deflects them. Figure 1.8 offers some practical advice on international communication.

Figure 1.8   International Communication Pointers

Even the relatively small business may have a global audience. Communicating across national borders has always been challenging. Now, however, the revolution in information transmission facilitates an almost instantaneous interchange of materials—business correspondence, institutional reports, and—for good or ill—artifacts of popular culture. Customers may come calling from formerly exotic places, confident that today's advanced shipping systems can get items to them quickly. The technical communicator must be aware of the needs of this new and growing clientele. Keep in mind some principles of this type of discourse:

- **Use "world English."** Your writing should be devoid of clichés, confusing references to sports, and slang.
- **Employ predictable word order.** Subject-verb-object patterns may be more easily translated, whereas inverted word order may not. Avoid passive constructions, vague indirect objects, and the assumed but unclear pronoun references of everyday speech.
- **Provide a translation in an alternative language.** For example, process instructions attached to merchandise sent to Canada should appear both in French and English. Some Internet sites maintained by businesses even offer the same text in both British English and American English. Being audience-friendly is a point of courtesy.
- **Supply a convenient, labor-saving means of return reply.** Use forms with blanks to fill in rather than imposing on your client the burden of creating extensive responses translated from the client's language to yours.
- **If constructing an Internet site that is interactive, follow the above principles.** Have your respondent communicate by supplying responses in a template that does the difficult work of sustained presentation for the client. Filling in small but critical parts and not worrying about having to supply text linking those parts will expedite communications for your client. Offering such templates in translation is also important.
- **Provide extra time for clarification of issues.** Do not assume that responding is easy for your client or that it will be for you.

Exercise

Revise the following sales blurb so that it is clear to readers for whom English is a second language.

> Howdy! Gotcha on the good side today? Let's start off on the right foot and look at one of the hottest new gizmos in hardware-crime deterrence—the double-loop iridium-alloy computer lock! Get one and boogie down!

Finally, *repetition* is a type of poor word choice that usually indicates failure to combine thoughts into complex or compound sentences—or to use parallelism. It lends to a statement the effect of being stuck in reverse. For example:

◆ In deed, if not in action; in thought, if not in belief.
◆ I'm qualified for this job because the job requirements are the same as the ones for my last job.
◆ That memo is the one the company is still having trouble with. It is just not communicating, is it?
◆ Really, we just need realistic goals.

Assignments

1. Revise the text of the following sales letter for clarity, brevity, and appropriate word choice. Names have been deleted and circumstances altered to protect the silly—nevertheless, this letter is based on a real example.

Dear Schmendrick Dealer,

HAVE YOU NOTICE?
—more toy-childrens commercials (going to get worse!)
—your mail box is full with holiday present catalogs
—department mall stores are fixed up and holiday-ized

IT'S CHRISTMAS!
    This year we have made to date together the BEST RETAIL holiday promotion. A GREAT LEADER, a $20. coupon, exciting copy, we list dealers, beautiful colors, and alot emotion. In edition to all of that, there is a bonus for yours. A salesperson from a partisipating store will win a trip for two to FLORIDIA!

HERE'S THE FACTS

Your store, a partisipating dealer in the HOME CIRCLE Holiday issue, which goes on sale December 1, 2004. There is guaranteed circulation of 22 plus million national readers.

You will get copy of the regional split from HOME CIRCLE with cover letter about a week before issue goes out.

All stores must be full-stock by Friday November 22. We need your cooperation to be sure that we can service you promptly, rapidly, and quickly. Call Schmendrick at 1-800-555-0202 for quick order placement. Lines more available in the earlier hours of the day.

1-800-555-0202

YOU need sales the best way to get them is to have STOCK. You need a good representation of Schmendrick on display available for hands on. We have features the A-1 and B-2 models. Again, consumers will bring in the $20. coupon, which is $20. off the list price.

NOTE NOTE NOTE NOTE!!!!!!!! The A-1 used as retail LEADER......BE PREPARED! The A is listed at $39.95 and net cost is $24.98. Again, remember, the A your advertising LEADER. You can't give the consumer a raincheck for the Holidays—they will buy from another, HAVE STOCK NOVEMBER 22.

Call me is have questions or I can help make your promotion better

Sincerely,

Beth Schmendrick

Beth Schmendrick
National Sales Manager

Remember that this document is a lightly fictionalized example based on a real sales letter. Put yourself in the role of a busy store manager preparing for the holidays—flooded with incoming mail, bombarded with requests, scheduling problems, and bills to pay. Ask yourself: How is this

letter supposed to be a piece of transactional writing? What should its goals be? Follow up by answering these questions:
   a. What strategic problems does the author encounter in trying to achieve those goals?
   b. How does the author solve—or fail to solve—these problems?
   c. Is there a structure at work in this letter?
   d. Is the content clear?
   e. Is the audience addressed effectively?
   f. What effects result from problems with grammar?
   g. Are *you* convinced? Would you respond to this letter?

Now, in small groups within the class, rewrite that letter so that it communicates. Be sure to review and consult this chapter for advice.

2. The memo below seems to have been composed after a three-martini lunch. Within small groups in class, revise its grammar and usage so that it makes sense. Then create a memo from the shipping department responding to Andros, and clarifying what has been done to remedy the problem. (Use the format of the memo below as a guide).

To: All Staff
From: Edmund Andros, Accounting VP
Date: 9 October, 2004
Re: What IS going on with outgoing invoicing?????

It has come to my attention that, as the saying goes, we have a no-win situation out there at our warehouse docks and at the shipping/receiving desk.

YOU—I'm talking to all of you who fill out the wrong forms over and over again make life unbearable for the folks who have to track merchandise.

UNDERSTAND—THE OUTGOING INVOICES HAVE THREE (3) PARTS. WHITE copy goes to the customer with the merchandise, YELLOW copy goes to Accounting. Pink copy only stays in shipping file you may need to look at it from time to time. Right?

Lets get it right now, do it for us all. Thankyou.

3. The following extract is taken from a complaint letter written by an irate customer. Revise it so that it will convince the reader without aggravating him or her. Then answer the questions that follow.

Dear Customer So-Called Service:

I have worked with all kinds but have never before seen such a bunch of crooks as your outfit. When I went to get a new Video Checkout card, the mental midget behind the counter asked for my Visa card and charged me a TWO DOLLAR FEE. The twerp told me that the fee would be REMOVED IN TWO WEEKS after I checked out a movie. So I'm supposed to pay interest on a stinking two dollars that the cretin charged me just to test my Visa? I told the pabulum-brained idiot that he could roast in Hades before I'd pay a cent. Then I got out of there. I was boiling mad for three weeks. Now I've cooled down just enough to write you a letter and tell you what's wrong with you.

Another thing. When I left, I did a little research on your outfit. It seems video rental is just the TIP OF THE ICEBERG. Your real money must be in real estate, movie production, and—you guessed it—copping A LOUSY TWO BUCKS each off of millions of folks like me. No outfit that makes two and a half BILLION dollars a year needs my TWO DOLLARS.

So you can take this letter which is worth exactly the postage I spent to mail it and SHOVE IT!

Yours for better business,

Harry Lasko

Harry Lasko

a. Does Harry—as he would say—have a "legitimate beef"? Why or why not?
b. Does there seem to be a customer-education problem that Harry's letter—in spite of its florid tone—implies? Define it and explain how you would solve it.
c. What changes can you make to the letter's structure and style to ensure that it will be considered, rather than discarded?
d. Imagine that you head the Customer Service Department. Can you write a calm, controlled response to Harry's letter?
e. As you write your response, what strategies of organization and style might you use?

4. Here is an inappropriate response to a customer's complaint—one that
   many customer service representatives might secretly love to write.
   Fix it, being sure to project an organized and caring tone.

Dear Mr. Numan:

I received your letter asking for warranty service on the
Dynamo electric guitar you bought your son this year. You
stated that the electronics didn't work right and you wanted
them replaced.

Mr. Numan, our warranty department looked at the remains
of the guitar which you butchered and then sent in. When that
instrument left the factory, it had a blue sparkle metallic polyester finish on it. No doubt you, your son, and his uncle Billy-
Bob had a fine time in the garage stripping all of our beautiful
finish work off of the guitar and replacing it with trashy
bleeding skull logos and a polyurethane finish that could only
have been applied by a house-paint brush and a palsied hand.

Of course, to add insult to injury, you clipped the wires off all
the pickups and tone controls so that you could remove the
wiring harness to get at the guitar's blue finish so that you
could dork it up. The big, dull blobs of cold solder gunking
everything back together incorrectly are your contribution
to fine instrument repair. The faulty hookups and the cold-
soldered joints—which you made—are the reason your son's
guitar now only makes a ground-lift buzz at 440 Hertz. If he
plays music like you fix things, this is probably a blessing.

I therefore refuse warranty coverage for the reason that the
customer destroyed the instrument. Maybe in the future we
should specify that our instruments are warranted only when
sold to those of a lineage Cro-Magnon or later.

Truly,

Roxette Jones

Roxette Jones
Service Manager

CHAPTER

2

Organizing Information

Introduction

There are many kinds of transactional documents—letters and memos (you just helped to revise some of them), short and long reports, résumés, and instructions. And of course, any of these forms may be sent electronically. Before you begin working on the formats of such documents and exploring ways of developing content within them, you must know the patterns of organizing writing. Some of the material presented here first hit the textbooks in the days of the ancient Romans and Greeks—some has arrived more recently.

Much of the strategy of organizing transactional writing parallels the technique of public speaking (Figure 2.1). The speaker must understand the intended

Figure 2.1 Parallels Between Writing and Speaking

Audience. Both writers and speakers must research their audiences, determining needs, motivations, and values.

Agenda. Based on the characteristics of the audience and the mission of the speaker or writer, the communicator sets the agenda—focusing the communication on a central point (thesis) or purpose.

Organization. The communicator selects a pattern of organization that will express ideas clearly.

Content. Detailed support and discussion must be developed within the organizational framework.

Transitions. Links between the parts of the organization must be built into the presentation.

Expression. Phrasing, mechanics, and diction become refined so that the audience will understand the communication.

audience, be clear about content and its organization, and use cues and transitions—just like a good writer. A good speaker knows, too, that certain types of
information are best presented in expected patterns, and this concept is also important for writers to employ, whether their goals are to inform or to persuade.

Patterns of Order

*The governing principles in organizing transactional writing are the use and the
recognition of patterns.* Certain patterns or templates of writing are psychologically satisfying; we expect them when viewing a clearly defined type of work.
For example, if we read a 1930s British murder-mystery, we expect an old
house, mysterious servants, strange noises in the night, a lost inheritance, a
woman in danger, and a clever detective. If we see a classic Gary Larson
cartoon, we know there will be beehive hairdos, wing-front glasses, talking
insects, mutated animals, and a bizarre punchline.

The patterns in business and technical writing are few but powerful. Let's
examine them, first looking at their origin and development.

Types of Order

The everyday world contains so much information that any transactional writer
must select from that vast storehouse to present a limited amount of data to
the reader. From the chaotic world, writers mold material into logical presentation. That can be difficult. For example, if you have been angered by rude or
incompetent service at a business and have tried immediately afterwards to compose a letter to the management, your anger may prevent you from being selective. You may start your letter by attempting to reproduce all the reality in a
frenzied, narrative style. Or the onrush of emotion may cause you to write as-
sociatively rather than logically—look at Harry Lasko's letter again (p.19) for an
example of this. But the reader of such a letter will not be inclined to help you—
or, indeed, to finish the letter—unless all that anger is contained and material is
selected to present in logical order. Such logical order has three main parts.

Transactional writing typically begins with a statement of claim—"I write to inform you of a customer-service problem in your company"—or a question
that will be answered—"Do you know that there is a serious customer-
relations problem in your company?" Claims may stress *factuality*, emphasizing that a certain event happened (example: the mechanic at the garage
caused damage to your vehicle at 9:10 a.m. last Friday). Some claims insist on
preference (example: the garage manager should prefer competent employees
and loyal customers). Many claims urge *action* (example: Flyby Oil Change
should pay for the repair of damage caused by the wrench dropped from the
hands of the clumsy mechanic).

*Once the claim is established, transactional writing then discusses details of the
problem.* Don't call the district manager an airhead for hiring the incompetent,

but describe how the service crew at FlyBy Oil Change dropped the wrench, tried to inflate your right front tire to fifty pounds pressure, and almost put transmission fluid into the oil.

Transactional writing then closes with a statement that establishes or encourages an obligation—not "No matter what you do, I'll never do business with your slimy outfit again" but "I request that your company cover the $450.00 damage to my vehicle, since that damage was caused by untrained employees of FlyBy. I expect your reply within the next fourteen days."

Essentials of Logical Order

As you have just seen, logical order involves a pattern similar to the thesis-development-conclusion approach you learned in English Composition. Putting the thesis at the beginning of the document results in a *deductive* approach that declares the point in the introduction, supports it in the body, and clinches it in the conclusion. Beginning with a question, following with data that must be understood, and concluding with an answering thesis constitutes the *inductive* approach. At its most basic, the support in the body might be description (discussion of appearance, geometrical arrangement in space, texture, or hue) or narration, which depicts an event moving through time.

There are differences, however, between basic essay writing and transactional writing. Transactional writing may not be unified by a *thesis* at all, but rather by a *statement of purpose*—especially if the document is a summary report:

> *Thesis:* There are four steps to take to register a student organization on campus.
> *Statement of Purpose:* This report will summarize the four steps required to register a student organization on campus.

You will see both focusing elements—thesis and statement of purpose—used in transactional writing. For example:

> There are four steps to registering a student organization on campus [Thesis].
> This manual will guide you through those procedures [Statement of Purpose].

Specific Templates

Governing the general logical pattern—which we can call for convenience

Message
Support
Closure

will be a specific template best suited for the job. Such templates include

◆ *Summary* (condenses a description of something, frequently narrating to do so)
◆ *Process* (shows how to perform a task or explains a procedure; is based on narrative)

◆ *Analysis* (divides something into its parts to discuss them—perhaps to evaluate them)
◆ *Comparison* (analyzes two or more different things)
◆ *Persuasion* (writes to convince the reader to believe or act)

Summary: Summary reports inform readers who need a "quick take" on an issue. The author of a summary must isolate the key elements of the material to be reported and state them in a coherent, organized fashion. Narration (discussion of events moving through time) and description (presenting spatial material in detail) may be used. Here is Julius Caesar's narrative summary of a military conquest: "I came, I saw, I conquered." Notice that *narrative summary* presents a sequence of events in time. Here is a small *descriptive summary*: "He is a short aristocrat with an exaggerated opinion of himself." No time sequence occurs in this example. Often, both narrative and description combine in summaries: "The haughty Caesar declared that he arrived and overcame."

Obviously, most summaries are longer than these brief ones; many can be quite intricate if what they depict is complex. You'll find a summary as a stand-alone report or as part of a longer document. A summary that precedes and describes a long report is often called an executive summary—it's produced for the general decision-maker in a firm who does not have time to look first at the details. Here's an example:

> The Wilco Products marketing strategy, designed to boost sales of garage door openers for do-it-yourselfers, did not book enough advertising in home improvement magazines to bolster sales. The following report will show five ways Wilco could target interested readership effectively.

Notice that this summary, which would appear before the main body of a report, contains both a thesis (first sentence) and a statement of purpose (the second sentence)—that common strategy of business and technical writing.

Another common summary is the condensation of a report or document other than the one you have written. Be sure not to duplicate the word order or the language of the original document and to identify the source of the document (more about this in the "research" section of this book). Here is an example of a summary of a longer report:

> The Sizemore report surveyed the need for vertical parking space at eight mid-sized hospitals in suburban Chicago. It calculated the square footage required to double the accommodation and analyzed the construction costs per square foot for representative structures. The report also compared the costs and benefits of vertical garages to those of conventional parking lots. Sizemore favored the vertical option.

Figure 2.2 shows a longer summary. Notice that the writer begins with a *message* explaining the focus of the research. Then, a detailed—but distilled—account of the project and sources follows: this constitutes the *support*

Figure 2.2   Summary of Student Research Interest

To:	Professor Clayton
From:	Karen Schuster
Date:	March 28, 2004
Re:	Project Studying the Feasibility of Day Care on Our Campus

Focus. Our college is experiencing a surge in enrollment, but the new students do not fit the demographic expectations typical of the past. Instead of being eighteen- or nineteen-year-old single, unattached freshmen, our new enrollees are older parents with obligations—either

* Single mothers needing credentialing to return to the workforce (average age, 25); or
* Couples seeking different careers after corporate downsizing (average age, 31).

These parents experience a major problem when our campus does not answer a basic need: what to do with the kids?

Description of Research. My research will assess the feasibility of establishing a campus-sponsored daycare center to serve students and faculty.

I will survey the Internet, CD-ROM databases, and current college guidebooks to determine how colleges solve this problem nationally—and to identify institutions that promote themselves as offering day care for the children of students and employees. Next, I will examine those colleges resembling ours to discover what models for such facilities will work in our region and with our mix of students. Finally, I will survey our students, faculty, and Student Services Committee to develop a proposed model for such a center at our campus.

Approval. Please let me know if you approve the project, and if you know of any leads I should follow.

section. Next, a *closing* statement requests agreement with and advice about the work. You will notice this format frequently in advertisements or instructions; a header provides focus, and the support follows. A closing statement then wraps things up.

Exercise

Summarize two kinds of material. First, write a one-to-three-paragraph summary of an event that has occurred to you. What do you need to leave out and what do you need to emphasize for the event to be clear? Second, summarize a long newspaper article. Test your summaries on your classmates—can another reader get the gist of what you have summarized?

Process: You made an acquaintance with process writing when you critiqued the video instructions in Figure 1.5. Process writing seeks to explain how to perform a task or how something works. It is strongly dependent on narrative time transitions such as "first," "next," "then," or "last." Some process explanations are terse operating instructions using imperative diction—such as this example:

CAUTION—IRRITANT—HANDLE ONLY WITH LATEX GLOVES!

◆ Remove epoxy stick from wrapper.
◆ Cut from stick just the amount required.
◆ Press and knead the inner and outer sections together.
◆ Apply to article when uniform color blend occurs.
◆ Wait for bond to harden in ten minutes.

Note the use of parallel structure in process explanation—each command begins with a verb in present tense: remove, cut, press, apply, wait.

Other types of process writing also use parallel structure and group tasks into steps—but these forms can be more complex. Such types include recipes, how-to books, in-house procedure manuals, sophisticated assembly instructions, and explanations of what happened. You will learn more about such writing later in this text.

Figure 2.3 displays a basic process document. Here, the Star Company has included on its package a set of directions for mounting wall bolts. Although these directions are simple, they begin with a focusing statement telling the reader that only three steps produce results. The procedures follow, and then—in place of a formal closing—the company's address appears. Its presence at the bottom of the text lets readers know that they can reach the company.

Exercise

Write simple process instructions for three common household tasks. Then check your directions by following your own procedures. What level of skill can you assume the reader will have? For example, if you write a recipe, can you assume the reader knows *why* to use butter instead of margarine when making caramel sauce? Can you be sure that the reader will be familiar with the process of caramelization itself, or do you need to describe it? Test the effectiveness of your instructions by having your classmates read them!

Figure 2.3   Process:
Installation Instructions
Source: Star, Mountainville,
NY, 1996

3 Quick &
Easy Steps:

1. Drill hole the same
size as Wallgrip®.
Tap into hole until
it is flush with
surface.

2. Turn screw with
screwdriver until
you feel anchor
expand and grip.

3. Remove screw.
Place fixture in
position. Replace
screw through
fixture and tighten.

2700-10

Figure 2.4 advises technicians about using a specialized piece of equipment—an edge-routing attachment for a Dremel tool. Since only a limited audience needs this accessory, the Stewart-MacDonald Company could reasonably assume its customers possess advanced knowledge. Nevertheless, the text provides a safety warning—always a good inclusion in any discussion about power equipment. Again, the text follows a message-support-closure pattern. It discusses the purpose of the device first, follows with a technical explanation supported by a figure, and closes with the safety alert. The company's name appears last, as in the previous example.

Exercise

Create a similar explanation for a specific audience already familiar with the specialized tools and techniques of some field.

Analysis: Just as process derives from the time sequencing of narrative, analysis derives from description—which doesn't need to be time-dependent at all. For example, if a "secret shopper" were hired by a restaurant chain to

Figure 2.4   Process: Use Instructions

Instructions for the Use of the No. 1613 Binding Router Attachment

This attachment enables you to rout the binding notch on the unfretted neck and curved resonator of a banjo or on the curved sides of a mandolin or guitar. It is designed to be attached to the Model 280 Dremel Moto-tool and is used with a Dremel No. 115 cutter bit. The attachment has been machined so that the long "nose" guides the cutter bit, channeling out a groove as shown in the photograph below.

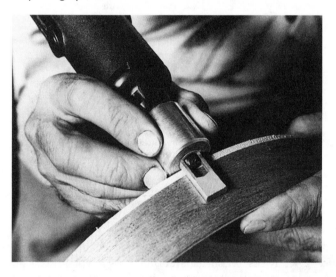

The attachment has been machined to provide two different cutting depths. The nose is slightly off-center and cutting depth is determined by the side of the nose held against the wood surface.

Height adjustments can be made by loosening the screw and sliding the attachment along the Dremel's front housing.

Practice cuts should always be made on scrap wood, inserting binding to check the desired depth. Minor adjustments in depth can be made by placing one or more layers of tape on the side of the nose being used.

Always wear safety goggles when using this tool and keep fingers away from the cutter bit!

STEWART-MACDONALD MFG. CO.

Source: Stewart-MacDonald Guitar Shop Supply Catalog, Athens, Ohio, 1981

report on Franchise #451, that shopper would evaluate that restaurant by isolating the restaurant's important features and reporting on them. The shopper would not provide the parent company with a minute-by-minute retelling of what she did in the restaurant and how she was treated, but might instead compose a report with six main headings:

Summary—On July 8, 2001, I investigated Franchise #451 because of complaints sent in to the home office. I found these claims justified.

Atmosphere—Restaurant was smoky. Floors needed mopping, and flies buzzed around the salad bar.

Service—I waited ten minutes for a regular hamburger, and when I requested ketchup, the clerk glared at me.

Food Quality—The burger was cold and red in the center. The mix on the soft-drink dispenser needed adjusting; my cola tasted watery.

Price—This restaurant has added ten percent to the price set by the franchise.

Recommendation—Based on my experience, I recommend an immediate and thorough review of this franchise unit by the district manager.

Notice that this report, brief as it is, contains a message or summary section at the beginning to communicate the essential point. It closes with a recommendation. In the support section, it classifies the features of the restaurant analytically under four main topics: atmosphere, service, food quality, and price.

Exercise

In small groups, try evaluating a local restaurant using a format similar to that discussed. Consider how you should determine criteria and demonstrate that the restaurant meets them. Then make a table or chart listing these criteria and providing space for notes to be taken while at the restaurant. When visiting the site to be reviewed, record your responses on your chart. Finally, construct a short report from your data, and present it to the rest of the class. Notice that analysis frequently proceeds from or is assisted by pictorial aids, such as charts, and that linking the parts of the oral presentation together is similar to the work needed to provide good written transitions, as well.

Reviews of cars, stereos, and other merchandise that appear in consumer magazines are also examples of analysis because they evaluate a product by assessing its components. And analysis may be used to sell. For example, Figure 2.5 describes parts of a brochure that categorizes the important attributes of new products, dividing the subject matter into sections.

Remember that classifying (grouping related items together), dividing (breaking something up into its components), and defining (showing what something is made of) are all analytical functions.

Figure 2.5   Common Elements in Print Advertising

Many longer advertisements in print, such as brochures, categorize the important attributes of products, dividing their presentations into sections corresponding to the elements of message-support-closure found in other examples of transactional writing.

Brochures often have *front matter*—text including a title—that declares a message, perhaps announcing a new item or urging its purchase. An accompanying picture demonstrates the product or the need for it.

Brochures often contain a *support* section incorporating vivid text with graphics in a sustained attempt to convince the reader of the merits of the article for sale. Often, pictures of the product function as "background" over which the explanatory text appears.

Brochures frequently *close* by requesting that the customer purchase the product. Contact information appears here—store locations, phone numbers, e-mail addresses, business hours.

Exercise

As a whole class, analyze a commercial product, first listing its many features. Then create a chart that categorizes the main headings governing discussion of the features, grouping related attributes together. Are there several ways to construct such a chart? What determines an effectively designed chart? Next, write—as a group—an analytical pamphlet presenting the product to consumers.

Comparison: When an analysis involves two different items measured against each other to see which one is better, then the analysis is called a comparison. A marketing report might compare two brands of frozen yogurt; a feasibility study might discuss the merits and demerits of two health-care plans; a pre-election pamphlet might provide analyses of two (or more) candidates' views. There are three types of comparison:

◆ *Segmented comparison* may compare each subsystem or category of items A and B; for example, the exterior styling of two different cars in the first paragraph, the engines in the second paragraph, and the safety options in the third paragraph. In technical writing, such comparisons often appear as tables or charts. Though segmented comparison is extremely useful, Figure 2.6 shows some problems that segmented comparison may create for the reader.

◆ *Holistic comparison,* used when the units are large and few, first discusses item A in its entirety, then analyzes item B; for example, the original Monte Carlo in a first paragraph—the recent Monte Carlo in a second paragraph. Figure 2.7 shows a holistic comparison in which Craftsman and Footprint handplanes are discussed separately, and the findings about them integrated. The opening summary of this text (a section from a much longer article) prepares the reader for the comparison, and the closing comments reiterate the author's viewpoint.

◆ *Likeness-difference comparison* shows how the two items are more alike than different, or more different than alike. If the items are alike in name only, the comparison would acknowledge that fact and merely explain their differences.

Figure 2.6   Comparison Pitfalls

> **Segmented comparison** may create problems for the reader. To avoid them, consider the necessity of *inclusiveness, equivalence,* and *transition.*
>
> **Inclusiveness.** If the parts are not large enough—inclusive enough—such presentations look scattered. Be sure to compare categories or systems, not unorganized lists of minute features.
>
> **Equivalence.** If the parts are not analogous or equivalent, no comparison can be made.
>
> **Transition.** The reader must perceive the internal logic of your presentation. Use transitions to link the segments together.

Exercise

Comparison is common in advertisements. Of course, for it to work effectively, a comparison pattern must be bolstered by relevant, specific detail. And it must be presented within either an inductive or deductive context. Figure 2.8 is an advertisement that compares two dog foods on the basis of digestibility and cost. Is the presentation inductive or deductive? Evaluate this advertisement—what is claimed in the large-type "message" section, and how is it supported in the text that runs down the left side of the page? How effective is the closing question in light of your analysis of the ad?

Exercise

Find a magazine ad similar to Figure 2.8 and critique it, asking yourself whether the article compares two items clearly and in detail.

Figure 2.7 Comparison: Review

Price vs. quality—Having examined the cheap handplanes, I was curious to see how the more expensive tools might compare. You'd think that spending twice as much money would get you a tool that works twice as well, but it ain't necessarily so. Case in point. While Sears' $27 smoothing plane was identical to cheap, hardware-store Stanleys, the English-made Craftsman jack plane I bought at Sears for $50 (catalog number 9-HT-37165) wasn't even remotely similar to high-end Stanley tools.

Despite a spiffy appearance, in fact, the Craftsman had a major, irreparable fault that rendered it just about inoperable: Its interior dimension was too narrow to allow proper lateral adjustment of the blade. And since the Craftsman's sole was ground out-of-parallel with the frog seat, it was practically impossible to set the blade to take an even shaving. You'd have to grind the blade's edge off-square to get it to work. On the one I bought, the lever cap was too wide, and its kidney-shaped hole had to be filed almost round before it would slip past the cap screw. And, like the No. 4 Sears I tested, the upper-end Craftsman had an inferior, single-step frog. My sample was also badly concave.

Although the Craftsman costs twice as much as a comparably sized Sears-brand plane, it's basically the same plane. All you get for the extra cash is a large brass depth knob, a spring under the lever cap and some chrome plating—hardly worth twice the price.

The Footprint—a high-quality, English-made jack plane—recently became available through U.S. tool catalogs. I bought mine for $43 from Woodcraft Supply (41 Atlantic Ave., Box 4000, Woburn, Mass. 01888). This plane felt well-made, and it was the only plane in my sampling that boasted wooden handles. But what else did I get for the extra money? The sole was flatter than most, but it still needed some work. The Footprint also had a two-step frog with ground seats, a cast Y-lever and a screw adjustment that moves the frog in precisely controllable increments—handy when setting up the plane. Overall, this plane had a quality feel equal to new, top-of-the-line Stanleys—a bit better feel than the Record planes I own.

I used to be something of a snob about my planes. I loved to restore noble old Stanleys, and I agreed with the Japanese that working with a fine tool is a spiritual experience. But, despite their shortcomings, the cheap planes I tested proved to be a practical alternative to spending top dollar for tools you plan to use only occasionally. It's clear that a little patience and tender tweaking can get these tools humming. Working with them convinced me that any tool that you've carefully tuned can become an extension of your hands and mind—even a $20 cheapie from K-Mart.∎

Source: Starr, Richard. "Tuning Handplanes," *Fine Woodworking,* No. 65, July/August 1987, page 56

Figure 2.8   Mock Comparison Advertisement

A Digestive Tract

What Makes **Barfo** Dog Food Better than **GreaseChunks**?

GreaseChunks: Slide down easy, but... tests by Canine Watch show doggie indigestion occurs quickly after eating. Result... a sad, dyspeptic puppy.

In contrast:
"**Barfo**'s mucilaginous texture facilitates digestion," says Canine Watch—and at an average of 20 cents less per pound, it's a better deal!

Your pup can eat better for less!!

Will **Barfo** bowl *you* over???

Persuasion: Persuasion is a connection between writing and oratory. Ancient rhetoricians contemplated how to keep people's attention during long persuasive speeches. For a presentation to be effective, said Roman rhetorician Quintilian, it must acknowledge the *audience, illustrate* the problem to be addressed, and state its *point*. Then it must *clear up* any misunderstandings about key terms or the path the presentation will take. Next, it must *present* the pros and cons of the issue. Finally, the presentation should *conclude* by supporting one position and asking the audience to do the same.

Look around you and you'll see this same pattern at work in transactional situations. For example, consider an appliance store. As soon as the customer walks in, he is greeted by a salesperson, who also determines why the customer is there. The salesperson listens to the customer's tale of woe—his microwave oven has shorted out, ruining his breakfast, and he never liked the way that oven worked anyway. So, the customer concludes, he needs a new, reliable oven with up-to-date features at an affordable price.

The salesperson explains those features to the customer; then, she shows him what her store offers in that price range. The customer likes what he sees—however, he remarks that a similar unit with a lower price can be found at MegaWorld Appliance down the street. But the salesperson then compares specifics; her model is clearly superior.

The conclusion of this presentation occurs when the cash register rings. Notice that the basic pattern might be outlined like this:

Message
    Address audience.
    Illustrate problem.
    State point (or thesis).

Support
    Define unclear terms.
    Examine pros and cons.

Closure
    Affirm position.
    Provide closure to discussion.

Clearly, this salesperson is a good coach, getting the customer to do her work for her and advising him of his options after he has clarified his stance. But the *persuasive writer* too must be a good coach. In persuasive business and technical documents—the ultimate transactional writing—the writer must make the work of reading a document seem easy and worthwhile. Headers, boldface, and underlined type encourage the reader to keep going; lists bordered by white space urge the subliminal message that "you can understand this—it's easy because it's broken down clearly for you."

Take a look at the short persuasive document in Figure 2.9. Here, a graphic initially attracts the reader's attention and literally serves as an illustration of the subject matter. The opening question addresses the reader and is followed in the first paragraph with a statement of the point. The second paragraph classifies the readership and explains that the magazine will appeal to all classifications. The third paragraph builds interest with detail, and the closing matter contains a special subscription offering. Notice that the magazine name and a guarantee of integrity appear at the bottom of the ad.

Figure 2.9   Persuasive Advertisement

SO YOU PLAY GUITAR? MAYBE YOU ALSO PLAY PIANO & KEYBOARD...

From the publishers of *Acoustic Guitar* and *Strings* Magazines comes *Piano & Keyboard* Magazine. No matter what the style of music you play, how often, or how well, you'll find *Piano & Keyboard* a welcome addition in your life. We've been working hard to make *Piano & Keyboard* a bright and engaging source of entertainment to all keyboard players, including you.

Whether you dabble in digital synthesizers or pound your living out on a concert grand, whether you rhapsodize at the harpsichord or re-mix with MIDI, *Piano & Keyboard* will enhance the learning process and playing experience for you.

In the coming year *Piano & Keyboard* will take you to the Garrick Ohlsson recital series in New York. To Australia for a visit with Michael Kieran Harvey. To France for the 150th anniversary of the birth of Gabriel Fauré. We'll look at Gospel music in digital arrangement. Some of the not-so-obvious benefits of the reproducing piano. Harpsichordist Gustav Leonhardt. You'll play the funky riffs of Lou Harrison. The sonatas of Padre Antonio Soler. And Brahms chorale preludes for organ and piano. In addition, you'll enjoy reviews of the latest editions, recordings, software, videos, and books. And then there are the informative reports about what lies on the horizon. Upcoming creative forces. And the names who make the news in today's global piano and keyboard scene.

We invite you to see for yourself all that *Piano & Keyboard* Magazine has to offer by returning the postage-paid card opposite. Or call us toll-free at (800) 827-6837. Or fax your postcard to us at (415) 485-0831. We'll rush you the latest issue of *Piano & Keyboard* and reserve your year's subscription (six issues) for just $19.95, <u>a savings of $9.75</u> off the single-copy rate.

. **One** final note ... you'll automatically receive our no-risk guarantee: If for any reason you decide that *Piano & Keyboard* is not for you, we'll gladly send you a full refund for the entire sub-scription price paid, no matter how many issues you've received. There is absolutely no risk to you.

ORDER NOW!
CALL TOLL-FREE (800) 827-6837,
OR MAIL THE POSTAGE-PAID CARD OPPOSITE,
OR FAX YOUR POSTCARD TO US AT (415) 485-0831 TODAY!

Source: Acoustic Guitar, March/April 1995, page 31
Courtesy of *Piano & Keyboard*

Exercise

How does the advertisement in Figure 2.9 demonstrate the principles of transactional writing that we have discussed so far? What techniques does it use to induce the reader to keep reading and to consider the offer it makes?

Ethics in Persuasion: Of course, not all advertisements adhere to ethical standards—since not all human conduct does. But your obligation as a technical writer demands creating ethical documents. Figure 2.10 discusses ethical problems in writing.

Figure 2.10   Ethical Concerns in Business Communication: Propaganda

The **"Golden Rule" truly applies to the principles of good practice in technical writing. Avoid language that is misleading, that omits important facts necessary for the recipient to make an informed decision, and that, in any other form, contains propaganda—or the attempt to persuade emotionally without a solid factual or logical basis. Below are some sleazy tactics that neither you nor your clients would appreciate.**

- **Name-calling.** Whether overt or by innuendo, this includes racial and ethnic stereotyping.
- **Transference.** An authority in one field is misused as an authority in another, or positive features of one thing are applied to an unrelated item. For example, a famous actor endorses a political cause. A beer can is colored red, white, and blue.
- **Pseudo-testimonial.** An unverifiable (or perhaps nonexistent) source testifies to the benefit of a product.
- **"Just like you."** The testimonial is given by someone pretending to share your aspirations, values, lifestyle, and concerns.
- **False cause.** Saying that A causes B if A precedes B. If one event precedes another, the first event may not cause the second. Both may be caused by a third event, both may be unassociated, or the early stages of B, unnoticed, might have caused A.
- **False cause.** Saying that A causes B if both A and B occur at the same time. Two events may or may not be related.
- **Loading the dice.** Presenting only part of the information the reader requires to make a decision.
- **Concluding about the whole from the part.** This is like reasoning that we can know all about an elephant just from its tail.
- **Coercion** or threat.
- **Bandwagon tactics.** Saying that "everybody's doing it."

Assignments ====================

1. Write a one-paragraph *summary* of
   - a college or institutional policy with which you are familiar
   - an article in your field of interest
   - this chapter of the textbook

2. Briefly explain the *process* involved in
   - accomplishing a particular home-improvement project
   - cooking a favorite recipe
   - completing an institutional task

3. *Analyze* your least favorite/most favorite business, evaluating each of its features using specific support.

4. *Compare two* different brands of the same kind of food (two different kinds of pizza, for example). Use three types of comparison, devoting a paragraph to each.

5. Write a short *persuasive presentation* that might be a letter to the editor urging a community project, a sales letter, or an outline for a persuasive oral report.

CHAPTER

3

Letters, Memos, E-mail, and Related Forms

Introduction

As you've seen, writing in the workplace is like and unlike the work required in composition courses. Certainly, both types of writing rely upon clear expression and organization.

But notice some differences, too. In composition and creative writing classes, the assignments frequently originate with oneself. Personal feelings and experiences become the basis for papers assigned to help the student grow comfortable with expression (grammar plus style) and organization. Yet transactional communication rarely emphasizes the writer's feelings. Instead, its focus is *other-oriented*. It responds to the questions: "What can I do for you?" "How will you benefit from accepting what I have to offer?" The real-world "assignments" in transactional writing are driven by practical needs within an agency, institution, or business.

In this and the following chapters, we will explore different forms of business and technical writing focused on the reader. Or perhaps we should say "readers"—because many such documents contain individual sections intended for different readers. The longer the document, and the more parts it has, the likelier this is. Such documents follow this pattern:

◆ *Message.* A beginning summary (sometimes titled "Summary" or "Executive Summary") may address a *project manager* who needs to know the whole picture.
◆ *Support.* Supporting sections follow (using one or more of the organizational strategies we've discussed)—which, because of their related but separate topics, can be *routed to different people* working on the appropriate stages of the project.
◆ *Closure.* A closing unit wraps up the findings, urges acceptance, or suggests further communication.

In many cases, the generalist manager reads the beginning and the end of a document, while scanning the headings of the body. The manager's subordinates read the pertinent sections thoroughly.

This approach to constructing documents influences the short external and internal correspondence that you will write. "External" correspondence, often the business letter, transfers information between one entity and another. "Internal" correspondence, often the memo, conveys information inside a single entity. Much informal communication in business, however, is accomplished by e-mail, fax, and various transmittals, which this chapter also will discuss.

Letters: The Basics

The message-support-closure plan influences the *business letters* you will write. It appears within the three forms of business letters:

◆ Full block
◆ Semi-block
◆ Simplified

Each of these forms is best suited for certain tasks, but all share certain protocols. Remember the following points about letters:

1. The salutation "Dear ——" is customary in business letters. It does NOT imply affection. A colon always follows such a salutation. If you do not know the name of the person to whom you write, address the salutation to the title or office: Dear Routing Manager. If you are unsure of whether the recipient is a Mr., Ms., or Mrs., use the person's title followed by the last name: Dear Director Brown.

2. Transitions should begin each paragraph so that the text of the letter reads smoothly. In essence, transactional writing always requires that you guide the reader through your thought process. Get familiar with such transition words as *first, next, then, in addition, in spite of, moreover,* and *however;* avoid nineteenth-century transitions such as *aforementioned, herewith,* and *heretofore.* Use *because* to indicate cause instead of *due to;* not "The bank refused to pay this check due to improper endorsement," but "The bank refused to pay this check because of improper endorsement."

3. The closing line usually consists of a word such as "Sincerely" or "Cordially" followed by a comma. It is NOT appropriate to vent your feelings in the closing, as in this example: "Yours for better business!"

4. Be sure your typed name appears after your signature. Below the typed name, put your title if appropriate.

5. If another has typed the letter for you, the typist's initials (in lowercase) are positioned in the bottom left of the letter. If you want your initials to precede these, they appear in uppercase: JS/pj, if Joan Smith wrote a letter typed by Phil Jones.

6. If the letter has a file name or number so that it can be retrieved on computer, this name or number follows the typist's initials on the next line.

7. BE SURE the letter is centered on the page so that it looks neat. No matter how short it is, a good letter should be the subject of several revisions so that what leaves your office is the best it can be.

Full Block Form

An example of *full block* form appears in Figure 3.1. Figure 3.2 presents a template or guide for all such letters. Note that because text is flush with the left-hand margin in this style, such letters convey a sense of directness.

Figure 3.1   Full Block Letter

123 Sunnybrook Rd.
Blanche, WV 25811
July 5, 2004

Loan Processing Center—Customer Service
Confounded National Bank
PO Box 33
Muddleston, WV 25125-1554

Dear Customer Service:

We have paid our first mortgage installment in two separate amounts because of an employee error.

Both payments were made before we received our mortgage payment book, which arrived after the initial due date. Since we had waited awhile for the book to arrive, we did not want to jeopardize our good standing with you. So we went to your branch office on 7/1, where a teller told us to "pay only the amount on the top line" of a form. The next day the teller called to explain that this amount did not include pmi/real estate taxes. So we returned to the bank and paid those. Both transactions were posted to Account Number 8XQZ90700, as follows: 7/1/04, by check #490, $661.38; 7/2/04, by check #491, $88.62.

To avoid future problems, we will remit the second payment noted in the loan book directly to your Muddleston office. If you have any questions about this transaction, please call us at 304-555-1239 or contact your facility in Blanche.

Sincerely,

Ray & Mary Lucas

Ray and Mary Lucas

Figure 3.2   Full Block Template

| Letterhead |

| Date of letter |

At least two spaces down—
More if needed to center letter

| Name and address of recipient |

Two spaces down

| Salutation (Dear—:) |

Two spaces down

| Paragraph single-spaced and flush with left margin |

One extra space between paragraphs

| Paragraph single-spaced and flush with left margin |

Two spaces down

| Closing line |

Three to four spaces down to accommodate signature

| Name and title, typed |

Two spaces down

| Initials of typist
File number |

Double- or single-space down

| Enclosure: "encl" if needed |

Double- or single-space down

| Copy information if necessary |

They seem businesslike and, at times, even bold. This is a good attribute— exploit it!

Features of full block letters include:

◆ Letterhead (name/address of sender)
◆ Date
◆ Name and address of recipient
◆ Salutation ("Dear so-and-so" is the norm)
◆ Single-spaced paragraphs with an extra space between them
◆ Closing line
◆ Signature
◆ Name and title, typed
◆ Typist's initials or writer's/typist's initials if needed
◆ Notations of file number, enclosure, copies sent

If you are not using stock company letterhead, make your own on the computer. Such letterhead typically contains the name and address of the entity sending the letter. Most programs provide templates that create letterhead after you supply the information. Typists generating personal correspondence may omit stating the name at the beginning, but should put the address above the date line in single-spaced form. If you know the mailbox number that follows the addressee's ZIP code, be sure to use it. Always remember that states need to be abbreviated in U.S. Postal Service designation. Not Az., but AZ.

Exercise

Figure 3.3 displays many of the common faults of poorly constructed letters. How many can you spot? Rewrite the letter so that it works effectively.

Semi-Block Form

This style resembles the full block form with the following differences:

◆ The date of the letter, the closing line, and the sender's typed name and title are aligned and tabbed in two-thirds of the way across the page.
◆ If you are not using letterhead, your sending address also aligns with the date, closing, and typed name.
◆ Paragraphs in the body have their first lines of text indented five spaces as if they were paragraphs in an essay.

A sample *semi-block* letter appears in Figure 3.4. While the full block format conveys a feeling of directness and immediacy, the semi-block form, because of certain resemblances to personal letter style, can seem more intimate. Many companies have sent letters of dismissal in semi-block form, however!

Figure 3.3 Problem Letter

Keller, Inc.
327 Brookdale Drive
Chesterfield, MO 63132

Joan Lupinski
24 Greenwood Ct.
Apt. A-4
Ballwin, MO 63212
March 6, 2004

Miss, Ms., or Mrs. Lupinski,

Regarding yours of February 4th: I'm sorry that you wasted your time and mine writing our company looking for a public-relations or personnel type job. Surely you must realize that with your limited experience, the most you could hope for would be a job sweeping metal shavings off the floor between shifts. Our janitorial jobs are all staffed right now. I'm directing my secretary to return your resume immediately herewith, since you're going to need it elsewhere when you look for a job.

I don't normally take the time to respond to each trivial job inquiry myself, but I was struck by the lousy way your aforementioned letter and resume presented itself. So I used a pen and corrected all of the preposterous spellings, the inflated claims, and the generally egotistical language. I guess that kind of language comes easily when your major in college is English Literature. So take note of my corrections attached and consider your experience tuition in the only school that matters, the school of hard knocks.

When you have some more experience under your belt, then start applying to companies who might use your training. But don't bother writing manufacturing firms like us.

Yours for realism in employment, I am

ALMALgm

Al "Bud" Malgum
President

Simplified Form

Some businesses prefer this style, which looks like a block format, but which does away with both salutation and closing lines. In place of the salutation appears a subject line explaining the content of the letter. This line is capitalized

Figure 3.4   Semi-Block Letter

123 Sunnybrook Rd.
Blanche, WV 25811
July 5, 2004

Loan Processing Center—Customer Service
Confounded National Bank
PO Box 33
Muddleston, WV 25125-1554

Dear Customer Service:

   We have paid our first mortgage installment in two separate amounts because of an employee error.

   Both payments were made before we received our mortgage payment book, which arrived after the initial due date. Since we had waited awhile for the book to arrive, we did not want to jeopardize our good standing with you. So we went to your branch office on 7/1, where a teller told us to "pay only the amount on the top line" of a form. The next day the teller called to explain that this amount did not include pmi/real estate taxes. So we returned to the bank and paid those. Both transactions were posted to Account Number 8XQZ90700, as follows: 7/1/04, by check #490, $661.38; 7/2/04, by check #491, $88.62.

   To avoid future problems, we will remit the second payment noted in the loan book directly to your Muddleston office. If you have any questions about this transaction, please call us at 304-555-1239 or contact your facility in Blanche.

                                        Sincerely,

                                        Ray & Mary Lucas

                                        Ray and Mary Lucas

entirely. The *simplified* style is halfway between a letter and a memorandum, which you will learn about shortly. It imparts a feeling of directness without a strong personal touch. An example follows in Figure 3.5.

Exercise

Dust off some correspondence that you have had to write to companies or institutions. Find a letter that you would like to revise; then

Figure 3.5 Simplified Version of Letter

123 Sunnybrook Rd.
Blanche, WV 25811
July 5, 2004

Loan Processing Center—Customer Service
Confounded National Bank
PO Box 33
Muddleston, WV 25125-1554

CUSTOMER CONFUSED BY BANK ERROR

We have paid our first mortgage installment in two separate amounts
because of an employee error.

Both payments were made before we received our mortgage payment book,
which arrived after the initial due date. Since we had waited awhile for the
book to arrive, we did not want to jeopardize our good standing with you.
So we went to your branch office on 7/1, where a teller told us to "pay only
the amount on the top line" of a form. The next day the teller called to
explain that this amount did not include pmi/real estate taxes. So we
returned to the bank and paid those. Both transactions were posted to
Account Number 8XQZ90700, as follows: 7/1/04, by check #490, $661.38;
7/2/04, by check #491, $88.62.

To avoid future problems, we will remit the second payment noted in the
loan book directly to your Muddleston office. If you have any questions about
this transaction, please call us at 304-555-1239 or contact your facility in
Blanche.

Ray & Mary Lucas
Ray and Mary Lucas
Lucas Contracting

cast the same letter into the three forms discussed above. What
differences in tone do you notice as the same content is expressed within three different styles? Next, write a letter responding to any of yours. What challenges do you face in
expressing your thoughts within the business letter format and
making them clear to another person or organization? Can you
see problems arising if letters were used to communicate information *inside* a single organization?

Memos: The Basics

The *memo* or *memorandum,* another basic unit of communication, is specifically designed to address the problems of communicating within an organization. Whereas the letter is normally sent from one institution or individual to another, memos are usually internal. They may be written on a company form (sometimes including a company logo), or a company style for them may be established by setting up a "macro" or template in the word-processing program in use. Common word-processing software often contains memo templates as well. All memos share certain features, however:

◆ They are headed by the word "Memo" or "Memorandum," which is centered at the top of the text.
◆ They identify date, recipient, sender, and subject.
◆ They work best if the text is flush with the left margin except for indented lists.
◆ Though they may have a copy line identifying other recipients, they seldom use salutation and closing phrases.

Two examples of memos appear in Figures 3.6 and 3.7; a suggested template for memos follows in Figure 3.8.

Exercises

1. Take the role of Vice President of Loan Operations at Confounded National Bank. You have just received the letter of clarification sent by Ray and Mary Lucas (the very letter of Figure 3.1). Respond to Ray and Mary Lucas—who seem very concerned about processing problems that simply shouldn't exist in your organization, but obviously do. Write a letter in either block or semi-block style to regain their goodwill. Test whether your letter communicates effectively by sharing it with a classmate for commentary.
2. As Vice President, write a memo to your staff detailing changes in procedures and accountability to ensure that this problem does not recur. Be sure to use headings and a message-support-closure strategy.

Varieties of Letters and Memos

Now that we have surveyed two basic types of short external and internal correspondence, letters and memos, we can explore their functions as they present neutral information, "good news" or positive material, and "bad news," or adverse content.

Figure 3.6   Sample Memo

MEMO

To: Ann Schmendrick
From: Ron Jones *RJ*
Date: July 3, 2004
Re: Workplace Handbook for Employees

To help us revise our employee manual, Hal Gustafson has acquired other company handbooks for our task force to examine. These are enclosed.

Most of what we've compiled already is similar in wording to the policy statements of other organizations; however, these other handbooks have:

- Better wording of sick-leave policies.
- Strong drug-free workplace statements.
- Clearer grievance procedures.

Please let me know what you think of the enclosed manuals. Call me at extension 949.

Thanks for your help!

Figure 3.7   Sample Memo

MEMO

TO: Angela Simms, Division Manager
FROM: Cal Brunton *CB*
DATE: August 27, 2004
RE: Employment, Plant Environment, and the ADA Conference

On September 20, QST-TV will broadcast a teleconference on the ADA and other legislation affecting hiring and accommodations practices. It is important that all supervisory personnel know how the current interpretation of these laws affects us.

Therefore, I have purchased viewing and taping rights for this program. Tapes will be available September 22, and may be checked out from Mary Lakewood at the Personnel Office. When returning the tape, please write in the log your preferred choice of time to discuss the conference and its significance to our facility. I have enclosed a copy of the conference's descriptive brochure.

Please let me know if you would like to schedule a speaker on the ADA within the next four months.

encl

Figure 3.8   Sample Memo Template

MEMO

TO: [Recipient]
FROM: [Sender] *Initial*
DATE: [Some prefer the date line to come first]
RE: [The word "SUBJECT" may be used instead of "RE"]

The first paragraph of a memo begins two spaces down from the subject line. Paragraphs are flush with the left margin and single-spaced. This "message" paragraph should be concise.

The second paragraph begins after a blank line space. Develop details in this and any following "support" paragraphs.

Closing commentary usually requests feedback.

encl [If there are enclosures, that can be indicated here.]

> This simple memorandum form becomes the foundation for several report styles, as we will see later. The basic parts of the short memo can be expanded to produce detailed ("articulated") sections of presentations.

Neutral, good news, and bad news letters provide the basic communication from an institution to an individual, or the reverse. Certain basic rules apply to all three kinds:

1. Proper grammar, punctuation, and usage are mandatory.
2. Letters must be block, semi-block, or simplified.
3. A letter should be consistent internally:
   • Address the same audience throughout.
   • Do not break format.
   • Use parallelism.
4. Enhance transitions with headers if necessary.
5. Use letterhead for the first page only.
   • Use plain paper for subsequent pages.
   • Number subsequent pages.
6. Put supplementary or distracting information in an enclosure.

Neutral Letters

Neutral letters convey information. They do not break bad news or inform the reader of a positive circumstance. If your role is merely informative, don't interject your feelings; the recipient of such a letter expects objective

Figure 3.9   Neutral or Informative Letter

Symonds Pastries
40 W. Walton St.
Chicago, IL 60610
312-555-5500

January 29, 2004

Dr. Elizabeth Stewart
4 Coburg Court
Oak Park, IL 60622

Dear Dr. Stewart:

Thank you for your inquiry about our catering charges.

We understand that you are eager to provide refreshments for your associate's retirement party. Enclosed are a list of our rates and options, and a description of our billing.

If you have additional questions, please do not hesitate to call us.

Sincerely,

Mary Symonds

Mary Symonds
President

encl

---

information. Just get to the point in the "message" section, and then supply the specific details in the "support" part. Figure 3.9 provides an example of a neutral letter. Much neutral correspondence summarizes material that the reader is anxious to know about—steps to take to avoid a problem.

Exercise

What has gone wrong with the text of the neutral letter below? Rewrite the letter so that it does its job.

Dear New-Home Purchaser:

I know I should have told you before now, but I guess that right after closing is a good time to give you the tax information you need. Your tax bill will come to the previous property owner this September, since you bought the house mid-cycle. He will probably send it to you. You need to take it to the Real Estate Tax Collection Office at the Duckville County Courthouse. They will rewrite the new bill and bill you. If you don't get the bill that way, just have the Collection Office make a new one. Since your bank has set up an escrow account to pay the property taxes, you need to take the new bill to the loan office of your bank. Don't call the Assessor's Office, because they don't do the billing. An assessor will be out to look at your property, however. Sometime over the next few months they are redoing the tax map for your part of the county. Since you have paid around what the currently assessed value is, they won't determine that an increase in payment is required. If it is, you can appeal to the Tax Board when you get your new bill. Hope you enjoy your new home!

Cordially,

Ron McGeever

Ron McGeever
Realty Associate

Good News Letters

Good news letters convey positive information. Don't be "coy" and withhold this information until late in the letter—get to it right away. Follow the basic information with details. It is best not to disguise a sales letter as a simple good news letter—the reader, expecting an immediate benefit, will become hostile when she finds there is a catch. Think of all the phony pitch letters you've tossed in the garbage; most of them probably had a beginning that seemed to provide you with an unrestricted benefit. Later on, the fine print disclosed an obligation. One pernicious variation of this pattern is the letter telling you that you've won free computer software; after two paragraphs, you discover that to claim the software, you must remit a "handling fee" greater than the software's cost. Figure 3.10 depicts a good news letter.

Exercise

Fix the text of the good news letter on the following page.

Figure 3.10 Good News Letter

Frank and Victor Medical Supply
32 Weir St.
Charleston, WV 25802
(304) 555-5678

July 14, 2004

Mr. Igor Calimari
65 Mountain View Drive
Transylvania, MO 61322

Dear Mr. Calimari:

I am pleased to inform you of your selection as an assistant to Doctor Victor. Your abilities will help him as he directs the completion of various research projects in human revivification.

Our firm offers a wide variety of benefits including company-paid health, life, and disability insurance. Supplemental insurance is provided through a participating agency's cafeteria plan. These, and our pension program, will be explained in full when you meet with Jamie Handel in Personnel.

We look forward to seeing you at 9:00 a.m., August 25, for your first orientation session. Welcome!

Sincerely,

E. Talbot

E. Talbot
Human Resources

Dear Occupant:

Hello, I'm your new landlord. I bet you're wondering if theres going to be a increase in rent. Let me tell you a little about myself. My name is Augie Batdorff, I bought this property with the money I inherited from my aunt. I need the income so I can go back to school and find a new career. Seems like hubcap replating is a dead-end these days, that's for sure.

Well, you probably see me around the place trimming those ugly trees or sealing the parking lot sometime this fall. Got to do it myself, those hired types really cost money!

I've been studying up all the leases and if the creek don't rise and the sun still shines, I think I can keep rent same for a year. Look for me around the place. I'll be wearing a red baseball cap. Thank you.

Bad News Letters

When writing one of these, knowing your audience is extremely helpful. If you are not familiar with your audience, role-play; put yourself in your reader's position. Use your message statement to say something good to the recipient—perhaps showing a benefit—while at the same time breaking the bad news. Don't prolong the reader's agony by burying negative information deep inside the support section. Use the closing section to keep future communication open. Figure 3.11 presents a *bad news letter*.

Exercise

Revise the text of this bad news letter so that it works.

Regina Bolt,

Your account with us is seriously past due. When we put the new dishwasher in for you you said you would pay us on the first of the month. I guess you figured you could pay us on the first of any month.

Since it is nine months now and no money received we have been carrying you longer than your mom did. We have sent you eight delinquent notices. Now we are fed up. We are going to give your account to a bill collector who will twist the money out of you someway.

Tandyn Ashcroft

This last example for revision points out one of the most aggravating types of letters to write concerning a problem: the collections letter. Though a detailed discussion of such letters is outside the scope of this book, you need to know how many organizations proceed. What follows is subject to

Figure 3.11 Bad News Letter

Frank and Victor Medical Supply
32 Weir St.
Charleston, WV 25802
(304) 555-5678

July 14, 2004

Mr. Igor Calimari
65 Mountain View Drive
Transylvania, MO 61322

Dear Mr. Calimari:

Thank you for your recent application for employment as assistant to Doctor Victor. We have reviewed your application carefully, and although your qualifications are strong, we have selected another candidate more experienced in nocturnal emergency organ transportation.

We will keep your resume on file for one year, and should another position open for which you are suited, we will certainly consider you again. Please send any updated credentials to Jamie Handel in Personnel.

We are certain that with your background, you will find the job just right for you. Good luck in your search.

Sincerely,

E. Talbot

E. Talbot
Human Resources

---

modification, as businesses frequently develop their own styles and approaches in sending these specialized bad news letters.

1. *Phase One.* When the bill becomes past due, the company mails a friendly reminder—sometimes with a postpaid envelope included to "make things convenient" for the customer. The text of such a letter stresses the valued relationship, mentions that the company must reconcile its records, asks the customer to notify the company if there is a disparity in records, and requests that a remittance be sent if there is no discrepancy.

2. *Phase Two.* The absence of customer response causes the company to mail a second, briefer notice reminding the customer of the obligation to pay—assumed now, since no correction of the bill was offered by the customer. At this stage, the company may enclose a payment booklet and a promissory note for the customer to sign "to continue in good standing."

3. *Phase Three.* Without a satisfactory response to the previous letter, the company writes the client stating that the unpaid balance will be turned over to a collection agency.

Note that hostile, confrontational, or illegal tactics are not permissible—the closing of the Ashcroft example (above) is not legitimate. Obviously, insulting someone hardly induces that person to pay. This brings up the importance of constructing transactional communication to get the desired results (in this case, payment) rather than pontificating in high-toned moral judgment—even though the tone might be justified and the temptation is there.

Good News, Bad News, and Neutral Memos

Within an organization, memos take the place of letters, but similar rules apply when communicating good, bad, or neutral news. Use the message section—or even the subject line—to sum up the issue, declare the good news, or state the bad news while offering a benefit. Use the support section to elaborate in detail. Close with a wrap-up paragraph that leaves the door open for future contact. Figures 3.12, 3.13, and 3.14 display examples of these memos.

Note that Mary Lakewood, the personnel director in Figure 3.14, might transmit a memo through her supervisor's office for wider distribution or

Figure 3.12   Good News Memo

MEMO

To: Sam Flint
From: Susan Jenkins *SJ*
Date: July 3, 2004
Re: Consulting Seminars

I'm pleased to fulfill your request to represent the company in our series of community mini-seminars on successful writing in the workplace. The three hours spent with high-school students should give them a better understanding of how writing is used in business, especially our firm.

Please stop by so that Stanley Astor can provide you with copies of the seminar scheduling. We want to make sure that we have honored your time preferences.

Thank you for your participation in this valuable program.

Figure 3.13   Bad News Memo

MEMO

To: All Employees
From: Susan Jenkins *SJ*
Date: October 3, 2004
Re: Quality-Control Workshop

It's time for our quarterly quality-control workshop, to be held in the cafeteria at 5:30 p.m. on Wednesday, October 12. The workshop should take three hours; all must attend.

Please stop by Stanley Astor's office so that he can receive confirmation of your attendance. Stanley will provide you with a packet of materials entitled *A Post-Deming World?* which you must read before the workshop. I'm sure you'll be surprised at what has happened to TQM and its spin-offs.

I hope you enjoy perusing the materials, and look forward to seeing you on October 12.

Figure 3.14   Neutral Memo

MEMO

TO: All Employees
FROM: Mary Lakewood, Personnel *ML*
DATE: December 3, 2004
RE: Supplementary Insurance Forms

Before the close of this fiscal year, we must fill out new payroll reduction forms for pretax payment of supplementary insurance.

The salary redirection forms are the same as those of last year, and should be completed after talking with our insurance representatives in the cafeteria from noon to three the week of December 9. Your form is enclosed.

Please call me at extension 925 if you have any questions.

Encl

---

greater authority. In many workplaces, such "through-memos" require a modified heading that would say:

TO:	All Employees
THROUGH:	Chris Sanderson, Vice President
FROM:	Mary Lakewood, Personnel
DATE:	December 3, 2004
RE:	Supplementary Insurance Forms

Such a memo would require two sets of initials—Sanderson's and Lakewood's.

As with letters, institutional style and format may deviate from these examples. In addition, some software programs place and space headings differently.

E-mail and Related Forms

You'll use several types of written communication related to business letters and memos. Some are discussed in the following sections.

E-mail

E-mail, or *electronic mail,* can be used internally in your organization or worldwide over the Internet. Basic memo format predominates here, though you will see exceptions. The particular e-mail template in which you enter the information normally resembles a memo, however. Conversational tone rather than formality often characterizes e-mail, though politeness is important. Recipients of e-mail dislike RECEIVING THE TEXT ENTIRELY IN CAPITAL LETTERS, WHICH SIMULATES THE EFFECT OF SOMEONE SHOUTING!!!! (In other words, don't do what I just did). Remember that communications sent via the Internet are assumed to be public—unless their access is restricted. Don't send anything through e-mail that you wouldn't want others—including your boss—to read. As with other business communication, proofreading is important if you wish to be understood and your ideas respected. For that reason, avoid using emotional icons such as smiley-faces, or colloquial abbreviations such as "U" for "you." Just as when writing conventional memos, indicate all recipients–don't use the "blind copy" function unless there is a compelling reason to do so (for example, a corporate security matter). And be sure not to mass-mail a communication directed at only a small part of your workgroup. The basic aspects of taste, clarity, and courtesy remain important in the electronic realm. Here is a sample e-mail message:

Subject: Sales videos

Date: Fri, 27 Mar 2004 17:01:21-0400

From: "Josh Anson"<hisemailaddress>

To: lindaholmes@heremailaddress

Hi, Linda! After the organizational meeting yesterday I developed a list of six videos helpful to beginners in sales. These would be valuable in training entry-level staff. When I find out current pricing, I'll zap the list to you.

In e-mail programs, the software template itself provides the space for appropriate information, typically filling in the day and time, the sender's electronic address, and the writer's name. Business e-mail text longer than a few lines benefits from dividing the communication into separate message, support, and closure parts. In addition, large documents referenced in the text should be denoted and sent as attachments. It is important to realize that although electronic transmission of information has proliferated, much business communication using electronic delivery is based on the standards of the memo, letter, and report.

Fax Cover Sheets

The sheet preceding a facsimile to be sent is called the *fax cover sheet*. The typical fax cover sheet derives its format from the memo pattern, though different formats exist. Be sure to type or print the information neatly, as the ability to reproduce characters distinctly varies with equipment. If you are printing the information by hand, use a black pen with dense ink. An example of a fax cover sheet follows in Figure 3.15.

Figure 3.15   A Fax Cover Sheet

FAX COVER SHEET
Schmendrick Electronics
40 Ridgeline Drive
Ace Industrial Park
Ellisville, MO 61301
USA
phone: 312-555-5550
fax: 312-555-5554

TO (FIRM):	FROM:
ATTENTION:	DATE:
FAX NUMBER:	FAX NUMBER:

STATUS (Circle One):

Urgent   Respond ASAP   Consider   FYI

Total Pages Including This Sheet:

Remarks:

Other Transmittals

Memo-like forms to be filled out and enclosed with business documents are frequently called *transmittals*. Such forms are precategorized, like fax forms, to save the reader and writer time. (An example of a retail version is the form/envelope used to send film to mail-order photo developers). If you need to design such a form, keep in mind that clarity is important. Figure 3.16 shows a type of transmittal. Figure 3.17 is a reminder about international communication.

Figure 3.16   A Type of Transmittal

Purchase Order

Schmendrick Electronics
40 Ridgeline Drive
Ace Industrial Park
Ellisville, MO 61301
USA
312-555-5550

DATE NEEDED _____

P.O. NUMBER_____

SHIP VIA_____

TO (FIRM):   ATTN:
ADDRESS: DATE:
TEL: FAX NUMBER:

TERMS (Circle One):

Open Net 30   30/60/90   C.O.D.   SPECIAL

Unit #	Quantity	Name	Unit Price	Total
			Subtotal	
			Shipping	
			Grand Total	

Figure 3.17 International Reminders

E-mail, faxes, and transmittals routinely travel around the world. Therefore, "world English"—and manners—need to be at their best in such correspondence.

- **Avoid localisms and slang.** Even though e-mail can be rife with such usage, employ the general English of broadcasters, not the extremes of formal or informal diction.
- **Use emotional icons sparingly.** :-) and :-(have achieved nearly universal recognition, though.
- **Provide straightforward order.** Use a subject-verb-object pattern, and be sure not to omit parts of speech.
- **Assist your presentation by supplying an alternative translation.** If possible, use preexisting templates with such translations.
- **Allow extra time for a response.**

Assignments

1. Imagine that the credit manager of Billings Heating and Air Conditioning—a company serving regional contractors—is confronted with a problem. Bayard Corey, a well-respected citizen of the town (and of a well-to-do family), has requested open-account status with net 60 days' billing in order to equip ten luxury condominiums he has begun building in a prestigious area. So what's the problem?

   Complication One. When performing a routine credit check on Corey, the credit manager discovers that Corey has delinquent balances exceeding five thousand dollars on each of two charge cards. The credit report also references a trail of unpaid bills to vendors connected with an out-of-state building venture gone wrong.

   Complication Two. Mr. Billings would really like the benefit of a sign advertising that the luxury condos are equipped with his units. In addition, the large sale would boost his finances and enable him to renegotiate his business loan coming up for review in four weeks. The sale of that many premium units would also delight his supplier, a manufacturer whose gloomy sales representative has begun seeking an additional dealer for products in the area to "pick up the slack" he feels Mr. Billings is not able to handle as an exclusive dealer.

   The class should divide into groups of approximately six each. Role-play within your group and divide up responsibilities after discussing the situation. Let one person write a memo or e-mail message from the *sales representative* to the manufacturer, Mega-Air, concerning whether the Corey sale is necessary for Billings to keep its exclusive franchise.

Someone else should create a memo from the *factory* to the sales rep. A group member representing a *factory official* must draft a letter to Billings—urging the sale?—while *Mr. Billings* and the *credit manager* will exchange memos reacting to the factory's answer. Finally, the credit manager must write a letter to Corey explaining whether to extend credit, or requesting further information on which to base the decision.

Be creative in your collaboration here—for example, the factory can offer extensions of credit or special terms to Mr. Billings to induce Billings to make the sale. Corey's problems can stem from a messy divorce and not from his own actions . . . use your imaginations! When everyone in the group has read and given feedback on all documents, then revise the documents into final form. Share with the class and evaluate each other's performance.

2. Now that you have completed Assignment 1, let's add another level of complexity. Assume that the manufacturer is owned by a parent organization, an international conglomerate with headquarters in Japan. A department in that company receives files of all correspondence related to all U.S. transactions, including those with Billings. Adjust the language of the documents sent to and from Mega-Air to facilitate international communication, and provide the resulting file with a cover letter or memo from Mega-Air to the parent company, K-R Corporation, explaining the Billings correspondence.

Our House to Yours: Using Summaries to Inform

Introduction

This chapter also covers transactional material that an organization sends externally to the public. Letters, memos, or transmittals themselves might introduce such summary documents as leaflets, flyers, bulletins, and announcements—all forms of communication relying on description and narration. To succeed, such summaries require continuity, clarity of expression, and an active voice.

Announcements and Bulletins

Look at the announcement below. List patches of unclear or wordy expression and rewrite the document, using clear language in a message-support-closure structure.

NEWS RELEASE—Waldorf High School PTA

RELEASE DATE: August 15, 2004
CONTACT: Leo Urquhardt, PTA President

PTA HOSTS SCHOOL FAIR

The school year is upon us, as we all know all too well. School will begin in just one short week. It was decided by the PTA and concerned parents who met in a long session that what we needed to have to stimulate parent and student interest and promote student welfare is a fair.

This fair will be held Saturday, August 23 (twenty-third) on the front grounds of the Waldorf High School. It has come to our attention that there is some question about deliveries of food. Those volunteers who will deliver food should arrive at the back entrance to the high school by the cafeteria. The fair will start at 6:30, with games, local musicians, food booths, and fun for all ages.

Dr. J. Harfield Nethercleft, Superintendent of Schools, will be on hand to answer any questions and to mingle with the crowd. He will be happy to discuss with you his new plan for split shifts which will begin next year.

All community members are invited to attend and to donate to the PTA funds. We need to buy science equipment for the school labs, replace the defective computers, and most importantly expand the parking lot for students by the new gym. Your contributions are tax-deductible.

Anyone needing to get more information should call me, the PTA President, at 555-3467.

#

Waldorf High School PTA
555-3467

As you read over this communication, did you notice problems with:

1. Imprecise language?
2. Continuity? (Does each part flow from the preceding part?)
3. Accuracy of description? (What will the fair be like?)
4. Clarity of narration? (What are the participants to do? Does the document accurately depict the event?)
5. Conciseness? (Can the great volume of prose be condensed?)
6. Audience awareness? (If this news release is intended for placement in the community section of a local newspaper or written to be read on a radio station's "bulletin board," will it target its audience effectively?)

Note that these issues are interrelated. For example, a document that uses imprecise diction will usually have trouble conveying ideas coherently; a document that is unsure of its audience will frequently express itself in vague language devoid of coherence. And remember the essential features of good summaries:

1. They isolate the main ideas to be communicated and place them in key positions in the text, either as topic sentences at the beginning of paragraphs, or as headers themselves.
2. They select from the mass of possible detail only the detail that best supports each main idea.
3. They proceed analytically, rather than following narrative continuously. This means that summaries generally do not "tell a story," but rather

extract the most important features from a potential story and present them without primary regard for time sequence. Thus, an essay test question such as "What are the four main causes of World War I?" should elicit an answer containing four main sections, each devoted to a cause—NOT a long narrative retelling all the events that incited that conflict.

4. They have the continuity achieved by making sure all supporting detail bolsters larger points, which in turn support one main concept to be made clear to a specific audience.
5. They have the clarity derived from proper word usage and appropriate grammar.
6. They are short. Freestanding summaries might consist of a short message paragraph, concise support, and a brief closing paragraph.

Exercise

Ask editorial questions about the public service announcements shown in Figures 4.1 and 4.2. What strategies work in them? What did they omit to their benefit? How are they successful transactional documents, encouraging the audience to pay attention?

Such "news releases" may take a variety of forms, and you may be called on to write one when your organization must communicate with the outside world via the mass media. Follow the general format of the examples depicted, keeping in mind a few pointers:

1. Double-space to make the announcement easier to read over the air at short notice. (Single-spacing of 14-point type makes an announcement easier to read silently). All releases for the media should be double-spaced in case they must be edited before use.
2. Use a modified-memo format to lend authority to the release. The word "more" enclosed by dashes (--more--) at the bottom of a page cues the reader to continue, while the ending "#" or "--30--" is a signal that the text ends. You might see news releases produced without either of these cues, however.
3. Provide newspapers, radio stations, and television stations with any announcement *at least* ten days in advance. A tidy text that is also punctual will more likely be read.
4. Call it what it is. If the release is truly for nonprofit purposes, then it may be called a PSA or "Public Service Announcement." This term has been misused greatly to include profit-making activities in hopes of getting free commercial time. Articles in business publications continually exhort their readers to disguise commercial enterprises as activities that altruistically benefit the public. Don't do this—it's not ethical.
5. Keep your announcement short—don't force someone else to edit your copy.

Figure 4.1   News Release: West Virginia Legislature (September 6, 1996)

RELEASE DATE: September 6, 1996
CONTACT: West Virginia Legislature

WEST VIRGINIA LEGISLATURE OFFERS
SPRING 1997 INTERNSHIPS

West Virginia college and university students are invited to participate in one of five internship programs offered by the West Virginia Legislature during the 1997 Spring Semester. With over 60 positions available, both undergraduate and graduate students are encouraged to apply for an invitation to a 60-day learning experience at the State Capitol in Charleston.

Two programs, the **Walter Rollins Scholars Program** and the **Robert W. Burk, Jr. Student Intern Program,** are designed for graduate students, while the **Judith Herndon Fellowship** and the **Frasure-Singleton Internship** are open to undergraduates. Both undergraduates (junior or senior) and graduate students are invited to apply for the **Legislative Information Internship.**

Because of the diversity of the programs, credit hours and stipends vary and must be arranged with each student's advisor.

Participants interested in gaining firsthand knowledge of the legislative process are encouraged to call the Legislature in Charleston at **1-800-642-8650** for more information on each program.

#
LEGISLATIVE REFERENCE AND INFORMATION CENTER
1-800-642-8650 (in state) or (304) 347-4836

Source: West Virginia Legislative Reference and Information Center

Exercise

Write a public service announcement or a news release for an organization to which you belong. Consider limiting the announcement to one paragraph of no more than five sentences. Be sure it is free of clunky expressions characterized by:

1. *Passive voice.* "The elections were decided by the committee" should become "The committee decided the elections."
2. *Clichés.* "A good time will be had by all," for example, is an ancient phrase!

Figure 4.2   Media Release

To:	Local News Media
From:	Baxter Environmental, Inc.
Date:	July 26, 2004
Re:	Local Environmental Assessment

Baxter Environmental, Inc. is a consulting firm hired in part by state grant money to evaluate environmental quality in your area and to suggest upgrading of local facilities.

Beginning August 2004 the Denham Township will be studied for adverse environmental impacts requiring remediation. Please expect to see our technicians inspecting sewers, storm drains, and sites along Bridge Creek.

Technicians will use harmless tracers in selected storm, commercial, and residential sewers to ascertain that these facilities work properly. Any personnel requiring access to residential or commercial property will carry identifying badges; visits will be scheduled neighborhood-by-neighborhood to minimize inconvenience.

This is an assessment visit only; no work will be performed at this time.

We thank you in advance for your assistance.

3. *Pompous elocution.* "It has come to our attention that . . ." is both clunky *and* self-aggrandizing.
4. *"Soft" verbs.* The weakest verbs in English are forms of "to have" and "to be." These forms both cause repetition and themselves repeat too often. Not "We want to have you write your congressional representative," but "Write your congressional representative!" Not "The project is divided into three parts," but "The project divides into three parts."

Double-space your release as if it were meant to be read aloud. Exchange these releases in your class; let others read your text back to you and suggest revisions for clarity.

Leaflets and Flyers

Often, an organization must distribute a summary brochure as a supplemental enclosure with a business letter, as a handout, or as information to pick up at a conference. These communications differ in format from the announcements examined above, though their goals might be similar. Such brochures might be single-sheet lists of features, letter- or memo-style explanations, developed descriptions printed on heavy paper stock, or tri-fold leaflets. In all cases, though, organizations need to consider the audience targeted and the context of the brochure. Will the readership be general or will it be a specialized clientele, familiar with equipment, procedures, and buzzwords? Will the handout need to stand alone, or will it be part of an integrated package, perhaps including visuals and explanatory data? What motivates the production of the document, and what is its goal? Here are some examples of this kind of writing:

1. A company encloses an informative explanation of its policies with its invoice and packing list. The document provides the customer service department address and number and specifies return procedures. Usually, such communications undergo periodic refining to keep them up-to-date and specific—thereby eliminating loopholes and unintentional vagueness.
2. An electric power provider includes a summary of safety tips with its bill. The summary contains an implied message, or subtext, that the utility company cares about the well-being of its customers.
3. A supplemental insurer produces an explanatory handout for its client's employees to pick up at a company meeting. The handout must be attractively designed so that people are motivated to take it, read it, and carry it around during a day of conferences.
4. A change in billing causes a long-distance service provider to insert a descriptive sheet in its mailings. A credit card company does the same. The summaries attempt to convey the subtext that each company still cares about its clients despite adopting unfamiliar procedures.
5. Insurance corporations, nonprofit societies, product distributors, and appliance manufacturers supply summary sheets to customers. These both inform clients about products they might want but don't have, and also explain the features of purchased products.

Figures 4.3 through 4.6 provide examples of such summary documents. Because the occasions for these summaries differ, their goals and audiences also differ. Some communications originally might have been part of larger packets of information.

As you read through these documents, take notes. Consider the following questions:

1. What is the setting in which such a flyer or brochure would be presented?
2. Who is the intended audience for the material? If there is more than one audience, please explain.
3. What is your holistic (overall) impression of each document? How does each communicate its message to the audience?
4. What editorial decisions might have been made as each item was produced? Why do you think so?
5. Can you discern a three-part structure of *message*, *support*, and *closure* in each document considered?

6. What parts of each communication do the work of the thesis, data, and conclusion sections of the traditional English paper?
7. Do you find that these traditional functions are explicit or "hidden" in the documents? Why?
8. What editorial decisions must have caused graphics to be used or not used? Why were those decisions made?
9. What strategies of transactional communication do these flyers and notices have which you might apply to other types of writing?
10. Is there a persuasive agenda at work in these documents, even though they appear to be informative?

Analyses of Documents

The following paragraphs use the material presented in Chapters 1–3 to provide short sample analyses of Figures 4.3 through 4.6. Study these paragraphs, and then write your own detailed reactions to these flyers or brochures as well as to four or five similar communications you've received recently.

Consumer Circuit

Because readers tend to overlook brochures and flyers that accompany monthly bills, this utility mailer uses splashy colors to invite people to peruse it. The "newsletter" style, complete with graphics, calls attention to the interest value of the text: boxed inserts, columns, and a bold banner headline are hard to ignore as they offer visual variety. Page two, however, uses a milder approach once the reader has been alerted; here, "feature" stories with complementary graphics capture attention and a "column" approach is not necessary. Sandwiched in the middle of the "safety" articles that convey concern for the consumer is a section extolling the benefits of a checkless payment plan—sure to benefit the company, and appearing also as a benefit for the consumer. Note the pictorial strategy in this newsletter. The "grounding" graphic on the first page counterbalances the "weight" of the big headline, while on page two, a large drawing closes the text. Even the recycling insignia in the lower right of this page carries a subtext of concern for customers and the environment.

The New AT&T Bill

This is benefit-centered writing. The tri-fold insert opens up to reveal a subdued graphic of the bill on the left and a corresponding list of featured benefits with topic headers on the right. A closing line at the bottom of this page invites response and provides a toll-free number—sending a message that "we care." The artwork is neither pushy nor neutral in conveying a similar message; both picture and text are persuasive entities, not just informative ones.

Figure 4.3 *Consumer Circuit* (August 1996)

AUGUST 1996

CONSUMER CIRCUIT

AEP AMERICAN ELECTRIC POWER
AEP: America's Energy Partner

LABOR DAY

OFFICE CLOSING

American Electric Power business offices will be closed Monday, September 2, in observance of the Labor Day holiday.

LIGHT UP THE NIGHT

Energy-efficient lighting doesn't end at the door! Use high-pressure sodium or mercury vapor lighting for the outdoor areas of your home. They provide more light, at a lower energy cost, and last longer than incandescent flood-lights or area lights.

HOW'S YOUR GROUND?

No, not the earth around your house but the grounding for your electrical system.

When your house was built, the electrical system should have been designed to meet the local building code or the current edition of the National Electric Code (N.E.C.). Has your electrical system been checked lately?

Just like the farmer who knows the quality of his yield depends on the condition of his ground, the quality of your electrical service depends on the quality of your electrical ground.

If your grounding system has deteriorated or is inadequate, you could be experiencing electrical shock, excessive burning out of light bulbs, or TV and radio interference.

If you think you may have a grounding problem, please contact a licensed electrician to inspect your grounding system.

Your grounding system could be deteriorating and you may not be aware of it!

Source: American Electric Power

Figure 4.3 *(continued)*

CHECKLESS PAYMENT PLAN

Imagine paying your electric bill each month without writing a check, mailing a payment or showing up in person to pay it. With the Checkless Payment Plan, you simply authorize your participating bank or financial institution to pay your electric bill. Each month your electric bill is sent directly to your bank and the amount owed is automatically deducted from your checking or savings account on the date the bill is due. You will continue to receive a copy of your bill from AEP each month indicating the amount and the date the funds will be transferred from your account.

For more information on this easy, convenient way to pay your electric bill, contact your local American Electric Power office.

WATCH OUT FOR OVERHEAD POWER LINES

Installing a TV, CB or ham radio antenna requires know-how and caution. Maintain a safe distance of at least 1 1/2 times the length of the antenna mast away from power lines in all directions.

If you're not sure of what you're doing, hire a professional to do the work for you!

Recycled
Recyclable

Figure 4.4 "The New AT&T Bill" (May 1996)

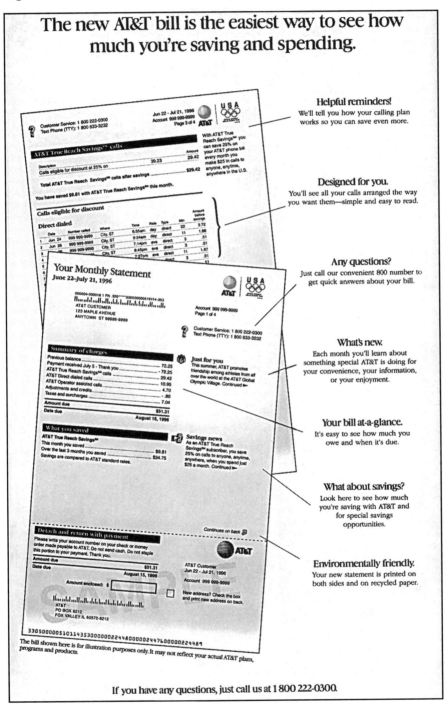

Source: AT&T

Figure 4.5   Finesse Finishing Papers & Sanding Blocks (September 1994)

FINESSE
Finishing Papers
& Sanding Blocks

"Finesse" finishing papers and blocks are made with evenly distributed grit particles of uniform size and shape. Finesse papers cut more quickly and smoothly than other papers, leaving no scratches when used properly. They last longer and do a much better job than ordinary finishing papers.

FINESSE SANDPAPERS:

• As with all "wet-or-dry" papers, including the 3M variety, soak Finesse paper in water overnight before using; it will cut better, with less "loading up." This is true for all wet-sanding lubricants, including naphtha, oil and mineral spirits. Finesse papers can be left in water for about five days before needing replacement.

• Most professionals prefer mineral spirits (paint thinner) or naphtha for wet-sanding lacquer finishes, and use water for synthetic polyurethanes and esters. Water can be used with caution when wet-sanding lacquer, but be very careful near exposed unfinished wood surfaces such as F-holes, volume and tone control holes, screw holes, neck pockets and routed pickup cavities. If water penetrates bare wood, splitting and lifting of adjacent lacquer can result; nor should exposed wood be soaked with mineral spirits or naphtha, although these lubricants are more forgiving.

• Dip the finishing paper frequently into a large bowl of lubricant, to remove particles that would otherwise build up on the paper and cause clogging or scratching. Use a piece of paper about two to four inches square. Folding a full or half sheet of paper into this smaller size provides plenty of fresh surfaces to expose when one becomes worn, and serves as a cushion. Professionals use foam rubber, felt, cork, wood or plexi-

glass as a "sanding block" for their finishing paper. Some sand a complete guitar with one or two four-inch pieces of paper; others switch paper more often. Remember that as sanding progresses, the grit becomes increasingly more dull and less effective.

• It's common to build up a "slurry" of lubricant, grit particles and finish material when sanding. Every minute or so, remove the slurry with a clean cloth.

• 500-grit paper is considered coarse by high-tech car finishers (the term "high-tech" refers to catalyzed poly-type automotive finishes; these are difficult to sand and rub-out, and require specialized equipment). Use 500-grit for quick sanding at the early stages, after the first few sealer coats of shellac, sanding sealer, primer or clear lacquer. In fact, you probably won't use ordinary 320 and 400-grit papers anymore. 500-grit is also excellent for fret leveling and final fret dressing.

• 800-grit can be used for the early stages of wet sanding. It cuts as easily as standard 600-grit, yet leaves no scratches. Instead, it leaves smooth sanding marks which are easily eliminated with grits of 1200 and up. 800-grit is also an excellent smoothing paper for fret leveling and polishing.

• 1200-grit leaves a surface on lacquer that buffs out with Mirror-Glaze #7 polish, skipping the usual Mirror-Glaze #4 polish or rubbing compounds completely. On the other hand, you may decide to use 1500 and 2000-grit for a more flawless finish. You can follow an 800 sanding with 1200, or if you've sprayed a smooth, flat finish, try wet-sanding with 1200 immediately, skip the 500 and 800-grits, and rub-out with Mirror-Glaze #7.

• Use 1500 and 2000-grits for ultra-delicate sanding. With 2000-grit, you can touch-up and spot-sand a lacquer finish, rub-out with #7 Mirror-Glaze, and leave hardly a trace that you were there. Of course, practice your technique on "scrap finishes" first.

• 1500 and 2000-grits are a must for synthetic finishes such as polyester and

polyurethane. Even basic 600, 800 and 1200-grit papers are a bit too coarse for a perfect rub-out on synthetics. (Ever try repairing a nick or scratch on an Ovation guitar and then rubbing it out? With 2000-grit or the Finesse sanding blocks, you'll be able to accomplish it!)

FINESSE BLOCKS:

• The manufacturer recommends that Finesse blocks be left in water, and that the water be changed occasionally to keep it clean. Keep a fresh bowl of lubricant handy to dip them in, as with finishing papers.

• You can use the blocks as substitutes for finishing paper, although this really isn't their purpose. They're much the same as sanding blocks wrapped with finishing paper, but much handier and more accurate.

• The blocks work well for touching-up delicate finishes. Try leveling and smoothing a lacquer or super-glue "drop-filled" chip or scratch using the blocks. We've found them superior for this work. The blocks are also great for ester and hardened Ovation style finishes.

• Finesse blocks will level runs and sags in much the same way you might use a short section of fine-tooth file. The small, flat rigid blocks can "rock" on the imperfection, letting you feel the high spot and allowing easy leveling. Follow the 400 and 1000-grit blocks with 1200, 1500 and 2000-grit Finesse papers when the defect is about 90 percent removed.

STEWART · MACDONALD'S
GUITAR SHOP SUPPLY

21 N. Shafer St. • Box 900
Athens, Ohio 45701 USA

1550 09/94

Source: Stewart-MacDonald's Guitar Shop Supply

Figure 4.6   1995 Annual Report to Policyholders

1995 ANNUAL REPORT TO POLICYHOLDERS

Net income for 1995 rebounded from year-earlier levels when reinsurance payments for our subsidiaries' claims

from the Northridge earthquake took their toll. As a result, funds for protection of State Farm Mutual policyholders increased $27 per policy and stood at $327 per policy at the end of 1995.

State Farm remains strong. The A.M. Best Co., which analyzes the stability and financial strength of insurance companies, continues to assign its highest rating (A++) to State Farm Mutual.

State Farm agents and employees remain dedicated to providing even better service to policyholders in the most cost-efficient way possible. A combination of superior service, competitive prices and financial strength is our goal. We made solid progress toward that goal in 1995 and expect even more in the current year.

Edward B Rust. Jr

Edward B. Rust Jr., President

Notice of Annual Meeting

The annual meeting of State Farm Mutual Automobile Insurance Co. is held each year at 10 a.m. on the second Monday of June at the company's Corporate Headquarters, One State Farm Plaza, Bloomington, Ill. All members may participate in the annual meeting and have a right to vote by proxy or in person. Proxies must be on file with the Corporate Secretary 30 days before the annual meeting.

State Farm Mutual
Automobile Insurance Company
Home Office: One State Farm Plaza
Bloomington, IL 61710-0001

153-9000.9

Figure 4.6 *(continued)*

STATE FARM MUTUAL
AUTOMOBILE INSURANCE COMPANY

Statement of Condition (In Millions of Dollars)

Assets	1995	1994
Cash and Short Term Investments	$ 488	$ 487
Bonds	25,555	24,287
Common & Preferred Stocks	15,891	12,123
Equity in Insurance Subsidiaries	6,205	5,596
Other Assets	6,617	6,349
Total	**$54,756**	**$48,842**
Liabilities		
Claims and Claim Expenses	$16,384	$16,078
Unearned Premiums	7,091	6,949
Other Liabilities	6,161	4,671
Policyholder Protection Funds		
Funds Assigned for Protection of Policyholders of Subsidiaries	$ 6,205	$ 5,596
Investment Fluctuation Reserve	6,634	4,271
Funds for Protection of State Farm Mutual Policyholders	12,281	11,277
Total	**$54,756**	**$48,842**

Summary of Operating Data (In Millions of Dollars)

	1995	1994
Premiums & Membership Fees	$24,141	$23,216
Less: Dollars for Claims	17,415	17,660
Expenses for Paying Claims	3,494	3,219
Service and Administrative Fees	4,453	3,990
Underwriting Gain or (Loss)	**(1,221)**	**(1,653)**
Plus: Net Investment Income	2,385	2,233
Other Income	102	149
Income Before Dividends and Taxes	1,266	729
Less: Dividends to Policyholders	1	206
Income Taxes	293	38
Net Income	**$ 972**	**$ 485**

The financial statements of the company are audited by an independent public accounting firm.

Finesse Finishing Papers

This specialized fact sheet addresses the needs of a restricted market, but also employs many of the strategies we've been exploring in this technical writing guidebook. A large, bolded caption provides initial focus, and main headings stand in bold capitals. Bulleted short paragraphs break up the text for ease of reading, and the layout leaves enough white space so that the reader is not repelled by a solid, unrelieved mass of words. This leaflet begins with a "thesis" section. It closes with the address of the supplier.

State Farm 1995 Annual Report

Recipients may resist reading this communication, since it accompanies a bill for insurance premiums. Such resistance may begin the moment the envelope is opened by a busy person, or the instant that an agent hands a leaflet to a customer. Here, the company has tried to maintain interest by organizing appearance carefully. Headings and boxed areas in the body of this text supply visual attractiveness and promote eye relief. The report uses a photograph of the company president and the facsimile of his signature to appear "personal," yet official, in announcing the latest news. Below the initial summary appears an invitation to the annual meeting of shareholders and the company's address—a type of closure implying stability and accessibility. The reverse side contains a financial statement followed by a list of the company's directors. Adhering to standard technical writing practice, this insert also uses short paragraphs rather than densely packed narrative. Such strategy encourages the reader's attention. Of course, this document has little pictorial interest compared to the glitzy, slick investors' reports that many companies use. This might be thought a disadvantage, but it is probable that the company wants to convey an image of stolid security without extravagance—few would pay an insurance company to be flamboyant.

Assignments ==

1. Check your mailbox, the packets of literature included with new appliances, and the leaflets received at work or at the booth in the front of a discount store. Then:
   a. Select two informative summaries from these or other sources and compare them, writing down similarities and differences. Which summary is better, and why? Bring the documents and the comparisons to class for discussion.
   b. Rewrite a chosen informative flyer as a news release, being sure to define the audience of such a release carefully.
   c. Next, write—either individually or in groups in class—a pamphlet containing an informative summary. Integrate graphics and color into your presentation. Compare and evaluate the brochures.

    d. If such a project is a group effort, your group could create a facsimile of an entire packet of material to be distributed by an organization. This packet might contain a news release, pictorial material, a fact sheet for "frequently asked questions," and a pamphlet.

    e. Find and report on an informational site on the Internet that consists of a "home page," providing a summary of services or an index of material—plus one or more sections referred to in such an index or summary. How is the presentation of summary information in cyberspace similar to—and different from—that of paper and print?

2. *Service-learning option.* If your class is affiliated with a service-learning program sponsored by your school, the class can perform the following activities:

    a. Help a social-service agency by designing an informative document for it. Perhaps this will be a guide to community recycling or a leaflet describing a women's resource center. You'll need to discuss carefully with your supervisor the intended audience and the requirements of the communication. Also, you'll need to study past examples in order to develop a proper frame of reference.

    b. Help a business by creating an informative brochure for it. Again, you will be working closely with a supervisor who will clarify the goals and requirements of the document.

If your institution does not have such a service-learning program, volunteer your talents outside of class. There are many possible beneficiaries of your informative writing skills—your employer or nonprofit organizations to which you belong, for example. The "real world" is an effective teacher!

CHAPTER

5

Directions and Instructions: Writing About Process

Process Explains "How"

Much writing done on the job—and much transactional writing—involves demonstrating how to do something, discussing how something was done, or showing how an event unfolded. This type of communication, based on narration or time sequence, is often called process writing. Process writing dominates the nonfiction book market as well; books illustrating the proper way to install a deck on a house, disclosing how to get along with others, revealing how to buy a car, or promising ten quick steps to slenderness all constitute examples of process writing. For our purposes, there are two general types of process explanations: descriptive and instructional.

Descriptive process writing discusses the principles behind an event or the fundamentals necessary to achieve an outcome. This description may be organized topically rather than chronologically if the components do not flow in a strictly linear sequence. For example, a report covering what buyers look for when purchasing a financially injured company may include a number of tasks which do not need to be performed in a set order. Descriptive process writing may also be used to explain the history of a policy or program—that is, how something came about. Such writing addresses those who need to be informed about an issue but who are not learning to enact a step-by-step procedure. Figure 5.1 shows an example of a simple set of suggestions that qualifies as a process description.

Instructional process writing (or direction writing) is task-specific. It shows the sequence of steps one *must* follow to achieve a goal. The steps must be taken in the prescribed order. Directions for mixing epoxy, recipes for bananas Foster, and charts depicting assembly of lawn tractors fall into this category. Figure 5.2 illustrates a set of instructions. This kind of writing looks easy to do, but the following exercise may reveal that it is not.

Figure 5.1   Description of Procedures

Weatherizing Your House for Winter

Before fall becomes winter, you should consider the following strategies to ensure that the cold stays out and the heat stays in!

- Check the condition of weatherstripping around all exterior door frames; replace defective weatherstripping with new.
- Replace defective door sweeps with new, flexible ones.
- Repair broken seals in windows.
- Replace worn caulking as needed.
- If you have a heat pump, clean and service it according to manufacturer's directions.

Exercise

Write a set of directions to perform an everyday task, such as backing up a car, cleaning a sink, paying for a burger, or pouring a bowl of cereal. Then read the directions to the class, which will attempt to follow the directions and pantomime the operation.

Process Writing Concerns

Audience

One of the biggest challenges with process writing, as you might have just discovered, is knowing what assumptions you can make about your readership. Does everyone in your audience understand how to drive a stick-shift? Has everyone reformatted a hard disk in a computer? Does everyone know what materials are necessary to caramelize brown sugar? Do your readers appreciate the need to heed safety warnings? It's important to brainstorm about one's audience and be able to anticipate your readers' needs. This is vital to prevent job-related accidents that might be caused by the failure to include safety warnings in instructions.

Figure 5.3 displays a safety alert embedded in a simple set of guidelines. (Note that this text, like that of a recipe or lab report, contains a "materials" section preceding the sequence of events to be followed). Figures 5.4 and 5.5 depict vivid warnings embedded in process instructions that also use clear graphics. Try to emulate such examples when designing process documents.

Figure 5.2   Instructions

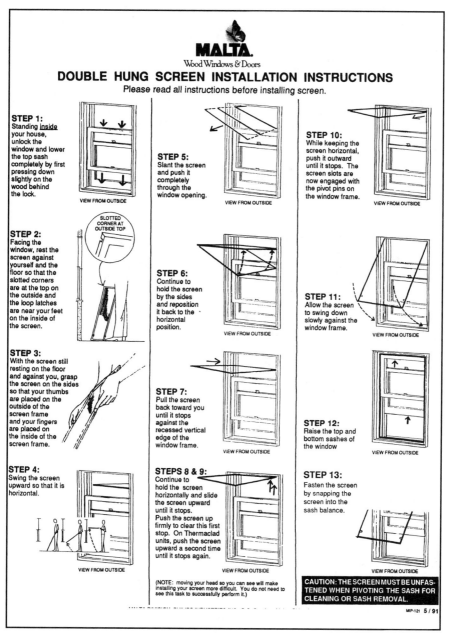

Source: Malta, Division of Tompkins Industries, May 1991

Figure 5.3 Safety Warning Embedded in Instructions

Instructions for Repairing Door Light Switch
On 1977 Ford LTD

READ ALL INSTRUCTIONS BEFORE ATTEMPTING REPAIR

Materials

Cyanoacrylate glue, latex gloves, safety glasses, tweezer, and small knife.

Procedure

Separate cracked halves of plunger.

Clean debris from surfaces to be glued; use knife if necessary but do not deform surfaces.

Select fresh cyanoacrylate glue.

CAUTION! Use care in operating glue tube. Cyanoacrylate glue bonds instantly to skin and causes skin irritation. Fumes cause eye irritation. Use latex gloves and safety glasses throughout procedure.

Hold one half of plunger with tweezers.

Spread thin bead of glue onto this half.

Place halves together; hold with tweezers for three minutes.

NOTE! Do not permit glue to cover electrical contact on switch.

Vocabulary

The language describing process must be familiar to your audience. Avoid specialized terms for a general readership.

Pattern

Figuring out a pattern of organization may be problematic as well. Do you need a list of materials preceding the instructions themselves? What if an alternate procedure exists?

Transition

How many little steps must you include to bridge the gap between the big steps—or can you assume a reader can get from one main step to the next? What guiding words—such as *first, next,* or *then*—will help the reader understand the process? Once you have arrived at the proper steps in sequence,

Figure 5.4   Instructions with Warning Labels

⚠ WARNING

This symbol identifies the most important safety messages in this manual. When you see this symbol, be alert to the possibility of serious bodily injury if the instructions are not followed. Be sure to read and carefully follow the messages that follow.

⚠ WARNING

It is the responsibility of assembler/owner to assemble, install and maintain gas grill. Do not let children operate or play near your grill. Failure to follow these instructions could result in serious personal injury and/or property damage.

SAFETY PRECAUTIONS...

Installation of grill must conform with local codes, or in absence of local codes, with **National Fuel Gas Code, NFPA 54 / ANSI Z223.1-**latest edition. Or **CAN/CGA-B149.2** Propane Installation code-latest edition. Handling of LP tanks must conform to **NFPA/ANSI 58**-latest edition.

If external electrical source is utilized with accessories (such as rotisserie), they must be electrically grounded in accordance with local codes or, in the absence of local codes, with the **National Electrical Code, ANSI / NFPA 70-1993** (or latest edition). Or **CSA Standard C22.2 No. 3,** Electrical Features of Fuel Burning Equipment (or latest edition).Keep any electrical cords and/or fuel supply hoses away from heated surfaces.

4 • Use & Care

Source: Char-Broil Use and Care Manual: LP Gas Grills, pages 4–5. Char-Broil, PO Box 1240, Columbus, GA 31902-1240

Figure 5.4 *(continued)*

⚠ WARNING

Grill is for outdoor use only. Grill should be operated in a well-ventilated space. Never operate in enclosed space, garage or building. Your grill is not intended to be installed in or on recreational vehicles and/or boats.

Do not install or use grill within 36" of combustible materials from back and sides of grill. Grill shall not be located under unprotected overhead (enclosed carport, garage, porch, patio) made of combustible construction.

⚠ WARNING

Carefully follow instructions in assembly manual and this booklet for proper assembly and gas leak testing of your grill. Do not use grill until leak checked. If leak is detected at anytime, it must be stopped and corrected before using grill further.

⚠ WARNING

When grill is not in use, turn off all burner valves and turn off tank valve.

Figure 5.5   Instructions with Warning Labels

INSTALLATION INSTRUCTIONS FOR YOUR NEW
FREE-STANDING
ELECTRIC RANGE

Before you begin - Read these instructions completely and carefully.

IMPORTANT - Save these instructions for local inspector's use.

IMPORTANT - OBSERVE ALL GOVERNING CODES AND ORDINANCES.

Note to Installer - Be sure to leave these instructions with the Consumer.

Note to Consumer - Keep these instructions with your Use and Care Book for future reference.

SAFETY

CAUTION:
For Personal Safety remove house fuse or open circuit breaker before beginning installation.

Be sure your appliance is properly installed and grounded by a qualified technician.

WARNING: To reduce the risk of tipping the appliance, the appliance must be secured by properly installed anti-tip device packed with the appliance.

All rough-in and spacing dimensions must be met for safe use of your range. Electricity to the range can be disconnected at the outlet without moving the range if the outlet is in the preferred location (remove lower drawer).

To reduce the risk of burns or fire when reaching over hot surface elements, cabinet storage space above the cooktop should be avoided. If cabinet storage space is to be provided above the cooktop, the risk can be reduced by installing a range hood that sticks out at least 5" beyond the front of the cabinets. Cabinets installed above a cooktop may be no deeper than 13".

 WARNING
• **ALL RANGES CAN TIP**
• **INJURY COULD RESULT**
• **INSTALL ANTI-TIP DEVICE PACKED WITH RANGE**
• **SEE INSTRUCTIONS**

STABILITY DEVICE
TOOLS NEEDED:

Phillips head screwdriver

1 3/8" open end or adjustable wrench

Bracket attaches to floor or wall to hold either right or left rear leg leveler. If fastening to floor, be sure that screws do not penetrate electrical wiring or plumbing. If this cannot be determined, use shorter screws that will not penetrate through flooring.

If the bracket came with your range, it is shipped attached to the lower range back. Remove and discard the shipping screw that holds the bracket and then follow instructions below.

1. Decide whether the bracket will be installed on the right or left side of range location.

SR10095
Pub. No. 31-10059
INT308-2

1

(Stability device instructions continued on page 2. . .)

Source: Installation Instructions, 1995, page 1. GE Free-Standing Electric Range Manual Pub. No. 31-10059. GE Appliances, Appliance Park, Louisville, KY 40225

then you must use clear transitions so that the reader will know how the elements in the sequence interrelate.

Exemplification

How detailed must you be? For example, will any hammer do or must you specify a $2^1/2$-pound drill hammer for the job?

Parallelism

Are the steps presented in parallel fashion to make them easier to comprehend? Be sure to

- *Use* active verbs of the same form
- *Begin* sentences with these verbs
- *Make* all sentences alike in grammatical construction
- *Study* this example!

Format

What format will best suit the presentation? Will it be:

- A *letter* to a client explaining the steps involved in establishing credit with your company?
- A *memo* to employees outlining the procedure for filing for the reimbursement of travel expenses?
- A *posting* on a job site reminding workers of safety precautions they must follow?
- A *brochure* enclosed with utility bills suggesting methods of conserving the use of gas and electricity?
- An *electronic* version of one of these forms?

Graphics

A picture may not be worth a thousand words, but it is nevertheless valuable. Observe that an illustration accompanies each step of the text of Figure 5.2 so that the reader cannot become confused. When you completed your in-class exercise, you probably missed the convenience of graphical aids as you attempted to explain the process you described. One important rule about transactional writing is to accompany the text with suitable illustrations *whenever possible*. This makes the material clear to the reader and provides visual relief. Imagine how a reader of the compact disc player manual (Figure 5.6) benefited from the clear graphics that help make sense of the text. (Be sure to read the appendix to this book for help with integrating visual material into your writing).

Figure 5.6　Instructions

COMPACT DISC OPERATION

4

5" (12 cm)　　　3" (8 cm)

Total number of tracks
Total playing time

10 4829

Play indicator

01 001

STOP

PAUSE

5

1　　　　　3

PAUSE
STOP

7　8　9

2　6

PAUSE
STOP 5
2 6
1

BALANCE　GRAPHIC EQUALIZER

LEFT　MAX　+10

RIGHT　MIN　-10

X-BASS　1kHz　10kHz

■ Care of compact discs

Compact discs are fairly resistant to damage, however mistracking can occur due to an accumulation of dirt on the disc surface.
Follow the guidelines below for maximum enjoyment from your CD collection and player.

● Do not write on either side of the disc, particularly the non-label side. Signals are read from the non-label side. Do not mark this surface.

● Keep your discs away from direct sunlight, heat, and excessive moisture.

● Always hold the CDs by the edges. Fingerprints, dirt, or water on the CDs can cause noise or mistracking. If a CD is dirty or does not play properly, clean it with a soft, dry cloth, wiping straight out from the center, along the radius.

■ Loading and playing CDs

1 Press the POWER switch to turn the power on.

2 Press the CD button.

3 Press the OPEN/CLOSE button to open the disc tray.

4 Place a disc on the tray, label side up.

5 Press the PLAY button to close the disc tray and begin playback from track 1.

● The "▶" indicator will appear and playback will begin.

● When a disc is loaded and the tray is closed, the total number of tracks and the total playing time will be displayed for several seconds.

6 Rotate the VOLUME control toward MAX to increase the volume, and toward MIN to decrease the volume.

Remote control operation
Press the VOLUME ∧ button to increase the volume and the VOLUME ∨ button to decrease the volume.

7 Move the BALANCE control toward LEFT to decrease the level of the right speaker, and move it toward RIGHT to decrease the level of the left speaker.

8 Move the X-BASS control toward MAX to emphasize bass sound. MIN represents a normal setting.

9 Move the GRAPHIC EQUALIZER control for any frequency toward +10 to boost the level for that frequency, and toward -10 to lower the level.

To interrupt playback:
Press the PAUSE button.
● The "II" indicator will light up.
Press the PLAY button to resume playback from the same point.

To stop playback:
Press the STOP button.

To switch the unit off after use:
Press the POWER switch to turn the power off.

5

Source: Manual INST 1749A. Sharp Electronics Corporation, 1992

Conversely, graphics can work to oppose the text if they are not carefully planned. For instance, the back side of a toggle-bolt package displaying the installation steps in reverse order—three pictures arranged right-to-left instead of left-to-right—may create confusion. It is the author's experience as a veteran of many home-improvement projects that such causes and such effects often occur. The following list gives guidelines for sequencing instructions properly.

♦ Don't assume specialized knowledge when addressing a general audience. Such an audience may need extra clarification of those transitions which you, the expert, find obvious.
♦ Clearly state any safety warnings.
♦ Use simple explanations to accompany illustrations.
♦ Make sure the illustrations accurately reflect the procedures described.
♦ Try beforehand the procedure which you will demonstrate—to ensure that you have included all steps in the correct order.

Invention

Concerns about pictorial and textual content as well as about format can be addressed in the planning stage early in your project. But often, whether you're writing process material by yourself or within the context of a small group, you initially confront two provoking questions: *How do I get started?* and *How do I keep going once I get started?* Considering all facets of a sequence and visualizing in pictures first (rather than thinking in words) may be part of your origination technique. Of course, the same writing technique will not work for every situation, so it is good to know many writing strategies. And though the forthcoming discussion of invention and development pertains to devising process instruction, its principles apply generally to the challenges inherent in producing writing of any kind.

Developing Ideas

Most of what follows derives from considering critical thinking. To write, you must think—you have to find out what's true, how things are connected to each other, and how to present these interconnections to the reader. In the first stage of a writing project, *begin with association*—that is, brainstorm for all kinds of ways to write about your subject. This activity should initially exclude nothing from consideration; keep searching for new possibilities (see the Invention Specifics section below).

Then, use *focusing*—in other words, once you have explored all the possibilities, narrow them down. What is your specific focus? What *one main point*—thesis or statement of purpose—best expresses that focus, or best tells the reader what your communication is about? Once you have that focus, which defines your agenda, think of examples that support your mission.

Finally, use reflection to develop ideas within a structure. At this point in your invention process, you should have a clear sense of your role. This will enable you to select a pattern of development. Analyze your needs here—for example, if you are providing instructions, what materials will your readers need to have to perform the task? What background understanding might they require? How can you link the steps together seamlessly so that the units of the task flow?

Invention Specifics

Here are some strategies to get you thinking about your writing. Try different ones!

Visualization Techniques: We think pictorially, and we may need to create a picture for others—either through words or through graphics assisting a document or an oral presentation.

◆ During your brainstorming phase, write down every idea on a big pad of paper; then go back to the paper, circle key ideas, and draw lines connecting them. Draw lines connecting subordinate ideas to main ideas as well. You'll end up with a web of interlinked concepts (Figure 5.7). Now you can assess emphasis; which ideas are most important? Which interconnections should you stress? Redraw the web on a fresh sheet of paper, and tinker with it until it suggests the proper order of ideas.

◆ During your planning phase, while analyzing your data and your mission, sketch out a storyboard (stick figures are okay) showing the ideas you wish to present in the sequence you will use (Figure 5.8). Any big gaps in the presentation will reveal themselves when visualizations of the steps are missing. The story line can split, of course, if two steps will occur simultaneously.

"Tasking" Techniques: On your computer—or on separate pieces of paper—type a set of tasks to fulfill regarding your topic (Figure 5.9). Leave ample room to fill in the answer below each heading:

◆ Describe the subject.
◆ Discuss the interrelationships within it.
◆ Explain what could be changed in your description without sacrificing the reader's understanding of the subject.
◆ Discuss how the subject fits into the larger requirements of your agenda.
◆ Note any problems your potential audience might have with unfamiliar terms or procedures.

Figure 5.7 A Sample Idea-Web—Used in Writing This Section of the Book

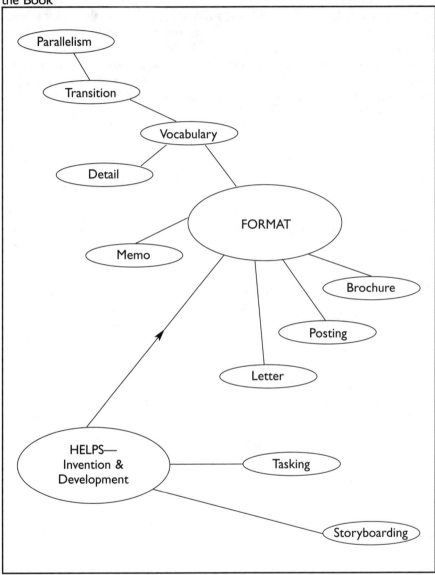

Using this task-list strategy may help you better define and describe your explanation so that it will be clear to your audience.

Once you have figured out your tasks and sequencing, you might turn the storyboard and task list into a flowchart combined with explanatory text

Figure 5.8 A Sample Storyboard

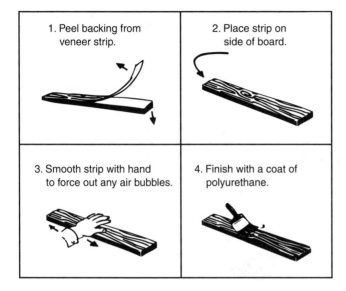

| 1. Peel backing from veneer strip. | 2. Place strip on side of board. |
| 3. Smooth strip with hand to force out any air bubbles. | 4. Finish with a coat of polyurethane. |

Figure 5.9 Tasking Sheets

| Describe the Subject | Discuss Interrelationships |
| Explain What Could Change | Discuss Larger Requirements |

| Note Problems |

Figure 5.10   Flowchart in Text

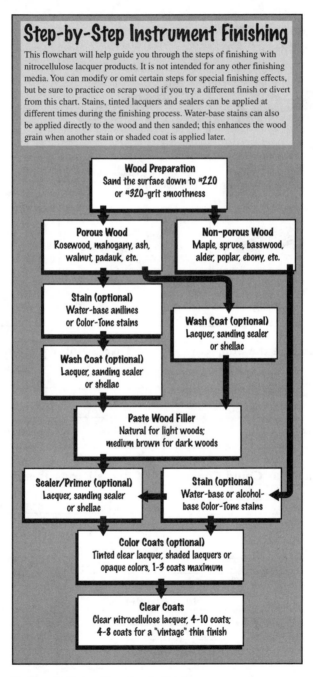

Source: Stewart-MacDonald Guitar Shop Supply Catalog

as in Figure 5.10. Such process descriptions can be integrated smoothly into the patterns reviewed in the next section.

General Patterns for Process Writing

The following patterns, used with much modification at times, govern most process writing.

For descriptive process:

Message
◆ Begin with a concise sentence or two about the process—either a thesis or a statement of purpose. A memo or e-mail might dedicate its subject line to this task—e.g., "Re: Wastewater disposal update." A simplified letter might use its subject line for the same purpose: "KEEPING ALGAE CONTENT DOWN IN PONDS." Or, if the description is a brochure, the title itself may state the main point.

Support
◆ Discuss the background of the issue or define unfamiliar terms.
◆ Describe the process clearly for your audience. Be sure to show the internal connections between any steps in the sequence.

Closure
◆ Close with a reference to your message, if appropriate. (Some descriptions, though, have only an implied closure and omit this part).

For instructional process:

Message
◆ State your message as a teacher would: for example, "This set of directions will help you assemble your new Willis Ten-Speed." The subject line of an e-mail, memo, or simplified letter may contain this message; often the title of an instructional sheet conveys the message rather than a sentence in the body.

Support
◆ List materials, tools, and safety equipment needed to accomplish the task.
◆ Explain the process clearly, using safety warnings as necessary. Employ parallelism throughout.

Closure
◆ Use a closing if appropriate—congratulate the reader on a mission accomplished.

Figure 5.11   Presenting Copy in Alternative Languages

Copy in alternative languages will convey your message across communication barriers. Consider the following helpful strategies as you design such copy.

1. Create *separate* versions, identical except for the text's translation into different languages. But supply all versions together. Or,

2. Create a *unified* translation by employing only one document and placing paragraph-by-paragraph translations consecutively in that document. And in either event,

3. Provide a *vocabulary list* before or after the translations so readers can see the exact equivalents of words—important if the subject is technical.

4. Accompany the text with *storyboard* pictures demonstrating the material.

International Communications

In a global marketplace, it's a good idea to provide process copy with clear graphics and with accompanying text in an alternative language. Study Figure 5.11 for suggestions about providing such helpful instructional material.

Assignment

Analyze the documents shown in Figures 5.12 through 5.14, noting intended readership, depth of content, and clarity of expression. What might you do to change the documents if you were to alter their audiences? If, for example, you were distributing the pickup installation instructions to a general audience, rather than a technically adept audience, what might you add? Discuss your findings with the class in small groups. Your group should then create instructional and descriptive explanations as posted directions, letters, memos, or brochures. As a service-learning option, your class might work with student government or the student services division of your college to create posted materials and informational leaflets.

Figure 5.12 Description

**Hallmark Gold Crown® Card Reward Certificate
Terms and Conditions**

1. This certificate can only be redeemed for regular-price Hallmark products at a U.S. Hallmark Gold Crown store.
2. When redeeming this certificate, please present your Gold Crown Card. This certificate is valid only for the person whose name appears on the face of the certificate and only through the stated expiration date.
3. Points cannot be earned on the face value of this certificate. Purchases made with this certificate are also ineligible for greeting card points and/or bonus points.
4. This certificate can only be redeemed for the exact amount shown on the front of this certificate at time of purchase. No cash value will be given. Customer is responsible for any applicable sales tax.
5. This certificate may be combined with other Gold Crown certificates. This certificate cannot be combined with any other offer and cannot be sold, reproduced, redeemed for cash or used for layaway purchases.
6. Hallmark reserves the right to alter, limit or modify program rules, regulations, rewards and reward levels or to terminate the Gold Crown Card program at any time at its discretion.

To all Hallmark Gold Crown retailers: Please redeem all certificates presented at your store. You will receive wholesale reimbursement on all certificates submitted.   w-ha 1010

Source: Mailer, 1996 Hallmark Cards, Inc.

Figure 5.13   Installation Instructions for Economy Piezo Pickup

1. Handle the pickup with care. Do not exert excessive force on the area where the wire joins the pickup as damage to the pickup may result.
2. The brass side of the pickup is the one that attaches to the instrument.
3. The economy piezo pickup is meant to be installed only once. Once it has been glued or taped into place any attempt to remove it will probably render it unusable a second time.
4. Try a "dry run" of the pickup installation so that you are sure of where you are putting it and to make sure that the wire goes where you want it to.

Adhesive options:
5. You may use a drop of 5 minute epoxy to bond the brass side of the pickup to the soundboard. Do not clamp the pickup in place as damage will result. Hold the pickup in place until the glue sets up.

   or:

   You may use a drop of C.A. instant adhesive to install the pickup. Use a **drop** and do not glue your finger to the pickup. I've done that. It is suggested that you wear a vinyl or rubber glove if installing with C.A.

   or:

   You may use a bit of strong double sided plastic tape if the soundboard is perfectly clean and free of dust and grime.
6. It is a good idea to support the lead out wire to avoid any excess strain on the pickup mounting or the solder connections to the pickup. Using clips or tape, start supporting the lead wire about 2 inches from the pickup.
7. For use in an acoustic guitar the best location that I have found for placement is about an inch behind the bridge plate.

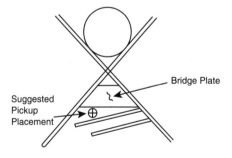

schatten design
124 Ottawa St. S. Kitchner, Ontario Canada N2G 3S9
(519) 742-3862 fax (519) 745-0953

Source: Insert, Schatten Design, 1997

Figure 5.14 Instructions

How to load your dishwasher

This is a random mixed load, the most common type you will have.

Make sure dishes are properly loaded to insure that water can reach the soiled surfaces. The wash arm in the bottom sprays water up. The tower that rises in the center sends water out over the dishes to wash the items in the upper rack. A spray arm in the top washes down also. Make sure to place tall items to the outside so the upper wash arm is not blocked.

WATCH OUT FOR THIS.

The wash tower rises through the center of the bottom rack during the wash and rinse portions of the cycle. Don't block it or load tall things next to it. Also, be careful not to let a portion of an item such as a pot or dish handle extend through the bottom rack. This could block the wash arm and cause poor washing results.

How to load the TOP RACK

The top rack is best for glasses, cups and saucers. Cups and glasses fit best along the sides. This is the place for dishwasher-safe plastics, too. Make sure small plastic items are lodged in tightly so they can't fall onto the Calrod® heating unit. Arrange stemware so that it cannot move easily. Don't let the glasses touch each other.

Sauce pans, mixing bowls and other small items may be placed—face down—in the top rack. The top rack is handy for all kinds of odd shapes.

8

Source: Use and Care Guide, Pub. No. 49-5491, page 8. General Electric Company, 1994

6

Using Analysis: Writing a Report

Subject Headings

Writing a short report involves organizing material topically and making the best possible use of internal headings. We saw earlier that main subject headings, which express the topics listed in outlines, can label paragraphs that follow them so that the reader can be guided along the path intended by the author. The use of subject headings inside reports is mandated by another aspect of transactional writing as well: different parts of a document are frequently read by different people. Whenever a document divides itself into topics or whenever a document clusters ideas together under a common heading to demonstrate their intrinsic relationship, that document is analytical.

Division

Division is the most apparent role of analysis. The report that uses division breaks up a complex subject for the reader, organizing itself according to the subtopics that together constitute the whole subject. Figure 6.1 shows a general pattern for such reports.

So, for example, a review of a new car model might discuss the engine in its first paragraph, the styling of the body in the second paragraph, and the safety features in the third paragraph. If the review were to compare the new model of the car with an older model, and if it adhered to the previous topical organization of engine, styling, and safety features, that review would be a type of comparison writing called *segmented comparison*, as shown in Figure 6.2. Comparison occurs when two or more things are analyzed together.

Obviously, for division to work, the parts of the whole must be significant. In planning a project that splits up a subject into subtopics, you need to ask yourself whether the main headings of your report are relatively equal in emphasis or whether some categories can be combined under larger headings for better organization. The reader of a technical document should perceive the outline—the structure—in the document's headings and subheadings (Figure 6.3).

Figure 6.1   Outline of Report Using Division

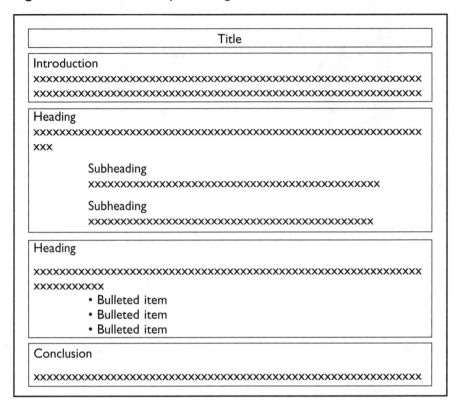

Figure 6.2   Segmented Comparison

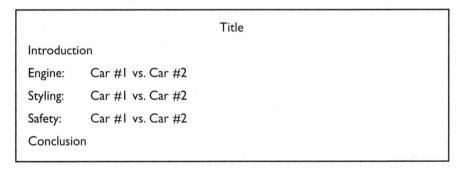

One advantage to the analytical report is its reader-friendliness. Because it uses headings, subheadings, and bulleted lists to divide issues, the report's format is more pleasing visually than that of a standard academic essay, the aesthetics of which do not often look inviting to the reader. Transactional writing must be read by busy people who like to have the argument of a report

Figure 6.3   Headings and Bullets

CAPITAL headings are for major sections (as in chapters of a report).

Main headings stand out in **bold**.

Subheadings may be <u>underlined</u> without bold.

- Bullets precede some subheadings.
- Bullets may be indicated in draft copy with an asterisk *.
- Bullets may be light or dark ○ □ ◆ ❑ ■ ❖ ●.

---

broken down for them; they do not like to view "bricks on a page," paragraphs unrelieved by headings and lists. Like graphics, the heading/list format encourages reader interest. This is especially noticeable when studying the connection between a speech and the same report text in print. Figure 6.4 is an example of a short report, originally a speech, which uses division but does not

Figure 6.4   Short Report Without Headings

Theater of West Virginia

Welcome to the Theater of West Virginia. I'm Kathy Holloway, a ranger for the National Park Service here at Grandview.

Grandview was originally developed as a state facility with the help of the Civilian Conservation Corps in 1939. Then in 1990 the National Park Service assumed responsibility for the site and it became a part of the New River Gorge National River. The history of the park and its name, however, go back to 1855 when Joseph Carper first farmed here and built his home in this beautiful area, which he called "Grand View" in honor of its magnificent vista.

Interpretation of the natural and cultural history of an area is an important objective of the National Park Service. The Theater of West Virginia, here at Cliffside Amphitheater, has been performing a portion of West Virginia history since 1961 with its production of *Honey in the Rock*. This story revolves around West Virginia's secession from the Confederacy and the many changes experienced by a fictional community during the turbulent years of the 1860s. A second production, *The Hatfields and the McCoys*, was added in the 1970s and is based on a true story of a blood feud between a West Virginia and a Kentucky family. The Theater of West Virginia excels at bringing to life some of the history and human drama of our beautiful state, and I hope that you will have the opportunity to attend not only tonight's show, but other productions as well. Please have a safe and enjoyable evening in your National Park.

Source: Holloway, Kathleen, Speech, 1996

Figure 6.5   Short Report with Headings

Theater of West Virginia

Introduction. Welcome to the Theater of West Virginia. I'm Kathy Holloway, a ranger for the National Park Service here at Grandview.

Grandview Park. Grandview was originally developed as a state facility with the help of the Civilian Conservation Corps in 1939. Then in 1990 the National Park Service assumed responsibility for the site and it became a part of the New River Gorge National River. The history of the park and its name, however, go back to 1855 when Joseph Carper first farmed here and built his home in this beautiful area, which he called "Grand View" in honor of its magnificent vista.

Natural and Cultural History. Interpretation of the natural and cultural history of an area is an important objective of the National Park Service. The Theater of West Virginia, here at Cliffside Amphitheater, has been performing a portion of West Virginia history since 1961 with its production of *Honey in the Rock*. This story revolves around West Virginia's secession from the Confederacy and the many changes experienced by a fictional community during the turbulent years of the 1860s. A second production, *The Hatfields and the McCoys,* was added in the 1970s and is based on a true story of a blood feud between a West Virginia and a Kentucky family. The Theater of West Virginia excels at bringing to life some of the history and human drama of our beautiful state, and I hope that you will have the opportunity to attend not only tonight's show, but other productions as well. Please have a safe and enjoyable evening in your National Park.

Source: Holloway, Kathleen, Speech, 1996

use headings; Figure 6.5 is the same text, but with headings added to keep a speaker or reader on track. *Notice that the reason papers in technical writing style use headings is because headings serve as markers in the text, just as pauses or changes in vocal inflection perform that function in oral delivery.*

Classification

Classification, or *clustering,* happens when a report must group separate kinds of data together to show that an internal relationship exists. Perhaps you have discovered that the seemingly unrelated observables x, y, and z are expressions of the same phenomenon, in which case your report would group them under the same heading. Classification thus assumes a general situation of which there are several specific cases. A comparison report disclosing that two politicians who claim to be different are really alike in

Figure 6.6 Comparison by Classification (Likeness-Difference):
A Simple Pattern

Title
Message: The two entities seem to be different but they really are alike.
Support: Support section discusses the resemblances:

• Resemblance 1
     • Resemblance 2
     • Resemblance 3

Closure: Reaffirms message.

their voting records, private statements, and attitudes has performed an act of classification. It has shown that despite apparent dissimilarities, an underlying sameness exists. Figure 6.6 depicts the pattern of such a report. Note that the closure must stress that the two politicians fall under the same *classification*.

Definition

Dividing a complex subject to explore its constituent parts or grouping related parts together to demonstrate similarity are often functions of *definition*. We tend to define by

◆ Illustrating through example (case studies, narratives, explanations of process)
◆ Explaining what something is made of (division)
◆ Discussing how something fits into a larger category (classification)
◆ Showing how something both resembles other items in its class and differs from them (classification/division or likeness/difference)

Use of Narrative

Definition, then, is itself frequently a species of analysis. And, just as in definition—where narrative must be subordinated to the purpose of explaining what something is—all major types of analytical writing must make narrative serve larger purposes. We tend to let "stories" take over from analysis in our everyday communication; for example, think of those times when you may have written an angry complaint letter to a company using chronological narrative instead of calmly explaining the problem feature by feature. Or remember your responses during in-class essay tests. A "C" answer might present the facts chronologically—as a story. An "A" answer might present the same data categorized according to the important main topics. When our emotions take

Figure 6.7 The Difference Between Narrative and Analysis

Narrative
- Tells a story.
- Uses chronological order.
- Does not necessarily emphasize important events (may include unimportant issues).

Example: "Once upon a time there were three bears..."

Analysis
- Divides or classifies based on topic.
- Is organized by topic, not by chronology.
- Groups subordinate data to support main points.
- Example:

Bear #1	Bear #2	Bear #3
Cool porridge	Warm porridge	Hot porridge
Small chair	Medium chair	Big chair
Tiny bed	Medium bed	Big bed

over, or when we are under duress, we tend to think narratively—we want to "tell a story." This universal impulse must be checked, however, when corresponding with a business or when constructing an analytical report. In analysis, narrative occupies the *subordinate* role of summarized exemplification. Figure 6.7 explains the difference between narrative and analysis.

What kinds of communication may best be presented analytically? Here are a few:

- Reviews of products, businesses, documents
- Assessments of effectiveness
- Discussions of background information
- Proposals for future work

In what kinds of format are you likely to write such reports? Such documents might be

- Letters
- Memos
- Semi-formal reports
- Formal reports
- E-mail or attachments to e-mail

Types of Reports

The following sections discuss the various types of reports usually encountered in business and technical writing situations.

Short Reports

In both letters and memos, the message section explains the purpose or thesis of the report. The support section divides and develops the subject of the report topic by topic. Finally, the closure section reaffirms the thesis or statement of purpose. Whether letters to clients explaining the advantages of new software or memos to employees discussing the features of a new voice-mail system, short analytical reports are common in business. As shown in Figures 6.8 and 6.9, they merely use the letter or memo shell to encapsulate information presented analytically.

Figure 6.8 Short Report in Letter Format

General Home Inspection, Inc.
1211 Rockville Road
Blair, MO 64321
312-555-5467

June 23, 2004

Thomas Ditson
435 Overview Drive
Creve Coeur, MO 63132

Dear Mr. Ditson:

We have performed a preliminary inspection of the property at Lot 25-C and almost everything appears to be in good order.

Drainage. The previous owner had a drainage problem—water standing in the front yard. A contractor installed drain tile to divert runoff successfully.

Construction. The house is 1 1/2 years old and meets or surpasses code in all respects except one. Current code does not permit incandescent lighting in closets. This house has five closets with standard bulb fixtures.

Site. The house lot abuts a county road on east; there are lots with houses both north and west. The south lot by the house is zoned residential only.

If you require more extended services from us, please do not hesitate to call.

Sincerely,

Bob Kusick

Bob Kusick
Inspector

Figure 6.9   Short Report in Memo Format

General Home Inspection, Inc.
1211 Rockville Road
Blair, MO 64321
312-555-5467

MEMO

TO: Site Crew
FROM: Bob Kusick
DATE: June 23, 2004
RE: Inspection of Lot 25-C

We must perform a preliminary inspection of the property at Lot 25-C. Our client is particularly concerned about the following points.

Drainage. The client states that the previous owner had a drainage problem—water standing in the front yard. We need to know what has been done to correct this.

Construction. The house is 1 1/2 years old. But does it meet current code in all respects? Please go over the house thoroughly to make sure.

Site. We need to check for zoning requirements, easements, rights-of-way.

Please attach all findings to this document, using it as a cover sheet. I will be meeting with our client next Tuesday.

Semi-Formal Reports

Semi-formal reports derive from the memo shell; they frequently have the "To-From-Date-Re" heading in the upper left corner, though the title "Memo" may be replaced with a title pertinent to the subject of the report. If such reports are presented as e-mail, the e-mail itself provides such a heading, and the e-mail subject line may state the title of the report. Often, though, a short e-mail will serve as a covering document introducing an attached report. In this case, the report should have a full heading. As do regular memos, short reports use single-spaced paragraphs with headings, and double spacing between paragraphs. Such reports can be written to inform or to persuade. As is the case with memos, such documents are customarily internal. Figure 6.10, for example, displays an internal report discussing the features of ISO 9000 international production standards. It must inform others in the company about the nature and significance of ISO 9000, and suggest a course of action for the organization. This report, then, both informs and persuades. Notice the "message" section's role as an executive summary.

Figure 6.10 "ISO 9000" Report by Bud Branch, 1994

Observe that this document—which is midway between a large memo and a more complicated technical report—retains the structural principles of a memo. It has the **heading** *of a memo, except that the subject line becomes the title of the report, and is centered. As does a memo, this document organizes the content by* **subtitles.** *A* **"message"** *section—here called "Summary of Report"—contains the gist of the communication. The* **"closure"** *section first sums up the details of the report, then states a recommendation. Such a document is typical of mid-sized written communication that originates in one department and circulates internally to others.*

Notice also that such documents, while retaining the standards of good written English, differ from academic writing in many respects:

- *They focus on a statement of purpose, not necessarily a thesis.*
- *They show their structure with clear headings—frequently questions.*
- *They avoid "intellectual," abstract language in favor of concrete terms.*
- *They do not necessarily use the terms "Works Cited," "References," or "Bibliography" to head a list of sources, nor is such a list always placed on a separate page as it is for academic essays.*

TO: Schmendrick Electronics Management
FROM: Bud Branch, Quality Assurance Group
DATE: December 9, 1994
RE:

ISO 9000

Summary of Report
This report explains the features of ISO 9000, surveys the positive and negative impacts of such certification, and recommends further analysis before our company adopts this potentially costly registration.

What Is ISO 9000?
In an increasingly global economy, the European market has become concerned about the goods and services available to it. Many companies—and countries—have had their own sets of standards for their respective products. In 1987 the European group International Organization of Standards (ISO) developed a set of guidelines for building a quality system adopted by the European Economic Committee; thus, the inauguration of ISO 9000 to warrant uniformity among products produced and marketed internationally.

Figure 6.10   *(continued)*

2

What Is Its Goal?

Experts believe that such uniform standards will generate security about the reliability of articles produced under their guidance. The burden of final quality control will be lifted from the recipient of such goods (Jordan 547).

Instead of having to meet several different audits and as many different quality standards, ISO 9000's goal is to have the world market comply with one set of global standards. Customers will know what they are getting when purchasing a product and/or service from an ISO 9000 certified company. This certification will be maintained by third-party organizations that have been certified by another third-party organization.

In addition, certified companies should incur increased efficiency in their operation through the reduction of waste and scrapped material. Also, improved product reliability should prevent customers from having to purchase as many spare parts and forestall excessive downtime.

The Three Types of Certification

ISO 9000 and 9004 are not the designations of certifications, but provide the principles for management of an organization in the development of its quality control manuals. It is through these manuals that the company will seek and maintain the following certification in one of the three areas listed. Diversified corporations will require different certification levels based on the product and/or service offered by different sites.

- ISO 9001 is the most comprehensive certification covering everything the company designs, develops, produces, and installs.
- ISO 9002 covers development, production, and installation only.
- ISO 9003 covers only final product testing. It does not include any procedures prior to the final testing before shipping. Because few companies obtained ISO 9003 certification in 1993, this certification may be discontinued.

What Is Life Like After ISO 9000 Certification?

One fear is the impact this program could have on a company and its operation. Auditing and assessment may take precedence over pleasing customers (Jordan 547).

Also, small companies with innovative designs may not be able to develop the procedures and follow-up required by ISO 9000.

Although pricing should not be affected in principle, the cost of registration is expensive—involving the development of procedures, documenting them, and auditing them. Once these systems are set up, savings are supposed to be gained through increased efficiency (fewer rejects, reworks, inspections). Although more customers and repeat business may offset the initial expenses, the customer should not necessarily expect price reductions as a result.

Figure 6.10 *(continued)*

Many ISO certified companies surveyed for this report privately state that[3] the systems they have developed in order to obtain certification are not fully in place after the registration audit. These organizations feel that at least two subsequent audits are needed before their documentation can be refined and understood by employees. Certification, then, becomes a process, not a prize awarded for a one-time inspection: the registrars come back every six months to follow up. The participants must understand the spirit of the standard in order to comply.

Because plans include product reliability, availability, and maintainability, the producer must furnish documentation ensuring these factors. But some companies state that the amount of documentation they must create stifles their flexibility, creates more work, and erects barricades to change. When a company overdocuments, it also must live with the inconvenient results, since what is declared in the paperwork must be accomplished (Mullin and Kiesche 34–40).

Conclusion

The articles and companies surveyed for this report affirm that if an organization obtains ISO 9000 registration and then loses it because of noncompliance, such would be worse for that company's image than being unregistered in the first place. Critics of ISO 9000 also argue that because it requires documentation only of procedures, it may ensure that a company with a mediocre level of quality will stagnate at that level. Of course, proponents believe recertification should be a continuing process of improvement of any company's operation and product. Yet many questions arise about the cost of certification and the reality of any potential payback.

Recommendation

Because of the program's newness, one cannot really perceive its ramifications. Our main competitor is not ISO certified, and does not directly compete with us in the same global markets. I recommend further research and much caution before we take the expensive step toward certification.

Print Sources Surveyed for this Report

"After ISO 9000." *Electronic Business Buyer* October 1993: 48–64.

Jordan, Jo Rita, "ISO 9000 and Analytical Instruments." *LG-GC* August 1993: 547.

Mullin, Rick, and Elizabeth S. Kiesche. "ISO 9000: Beyond Registration." *Chemicalweek* April 20, 1993: 34–40.

Rabbitt, John T., and Peter A. Bergh. *The ISO 9000 Book: A Global Competitor's Guide to Compliance and Certification.* White Plains, NY: Quality, 1993.

Sanders and Associates. *How to Qualify for ISO 9000.* New York: Sanders/ American Management, 1993.

Proposals

A proposal intending to persuade may be written as a semi-formal report. The material covered in a proposal's headings typically appears as follows:

◆ The *Message* section is called "Summary," "Introductory Summary," or "Introduction"; this section must summarize the purpose of the report and state its thesis or focus.
◆ The *Support* section contains the following headings:
   ◆ One or more content-related headings such as "Problem," "Overview," "Area to Be Investigated," "Field of Research," or "Issues to Be Addressed."
   ◆ One or more solution-related headings such as "Solutions," "Proposed Plan," or "Description of Proposed Project."
   ◆ A heading covering supporting materials, which may be called "Documentation," "Research," "Sources," or "References."
   ◆ A heading called "Timetable," "Estimated Time of Completion," or "Schedule."
◆ The *Closing* section may be headed "Request for Permission" or "Approval."

The proposal may shift or modify these headings as necessary to appeal to a specific audience. For example, a "References" heading might appear at the end of the proposal above a list of clients who would recommend the author for a project. Or a grant proposal might use a "References" section to list a bibliography. Figures 6.11, 6.12, and 6.13 depict examples of such proposals. Notice the variation in headings but the overall similarity in structure.

Analysis of Proposal.   Figure 6.11 is a proposal for a research project by Angela Stewart ("prospectus" is another word for proposal). Angela uses a memo shell, but replaces "MEMO" with "RESEARCH PROSPECTUS"—thereby labeling the document more specifically. Her message section, the "Introductory Summary," explains the purpose or agenda of her report: it will inform the reader and demonstrate solutions. Angela has provided an outline of her paper-to-be in the next section, using bold headings to accent immediately important information. She then explains her strategy of research, and lists a number of useful sources—not in strict MLA style, since we had just begun to work with bibliographical format in class (we'll consider such format in the next chapter). She concludes her support section with an estimated time of completion, and closes the paper with a request for my approval.

The research proposal is a common college assignment—useful because it can become the foundation for an extended paper. Nonetheless, note how such a document might not emerge from an academic context at all—if Angela were writing a proposal to build a deck on the back of my house, that prospectus would be similar. It would probably be organized like this:

◆ Summary
◆ Types of deck design that would be feasible

- Methods of construction
- Estimated time and budget of project
- Closing requesting action

Such a well-written proposal would induce me to get started on that deck immediately. Notice how the content may change—there's a big difference between a paper proposal and a deck proposal—but the method of organization remains similar.

Figure 6.11   Proposal in Memo-Shell Format

RESEARCH PROSPECTUS

DATE: July 31, 1996
TO: Dr. Brian Holloway
FROM: Angela Stewart *AS*
SUBJECT: Ocean and Sea Pollution

Introductory Summary
I plan to complete a multi-sectioned report on ocean and sea pollution. The primary purpose of my report is to inform the reader of the problem; also, I hope to leave the reader with a sense of responsibility and show how each individual can make a difference.

Areas to Be Studied
My research will involve the following areas:

- **Ocean and Sea Pollution Defined**
  The first section will answer the question: What is ocean and sea pollution? I will clearly explain this type of pollution and how devastating and widespread it has become.
- **Causes of Ocean and Sea Pollution**
  In the second section, I will examine the most prevalent reasons for ocean and sea pollution.
- **Effects of Ocean and Sea Pollution**
  The third section will define the tremendous impact that contamination has on the ocean environment.
- **Positive Ways to Control or Alleviate Ocean and Sea Pollution**
  Finally, I will present the most effective ways of controlling further pollution. I will also share what I view to be the best solution and what the reader can do to help.

Methods of Research
I located several books at the Learning Resource Center, but none were specifically about ocean and sea pollution. There were quite a few books on water pollution; however, all of these were either no longer current or too technical for my intended audience.

Figure 6.11   *(continued)*

2

The information I found at the County Public Library was very outdated. It may, however, be beneficial in comparing past contamination to that of today. The four books that I found are:

1. Bartlett, Jonathan, ed. *The Reference Shelf: The Ocean Environment.* Vol. 48. No. 6. New York: Wilson, 1977.
2. Loftas, Tony. *The Last Resource: Man's Exploitation of the Oceans.* Chicago: Regnery, 1970.
3. Marx, Wesley. *The Oceans: Our Last Resource.* San Francisco: Sierra, 1981.
4. Moorcraft, Colin. *Must the Seas Die?* Boston: Gambit, 1973.

ProQuest has been my most useful resource. Using this system, I immediately found several germane articles. The titles are:

1. Baker, Beth. "Nations Coming to Agreement That Polluted Oceans Need a Cleanup." *Bioscience* Mar. 1996: 183. Access NO: 02732573 ProQuest Periodical Abstracts-Research I.
2. Brancatelli, Joe. "What Is Polluting Our Beaches?" *Home Mechanix* Mar. 1995: 44–45. Access NO: 02276392 ProQuest Periodical Abstracts-Research I.
3. Heinrich, Katherine M. "New Study Traces Padre Island Trash to Shrimpers." *National Parks* Sep. 1995: 21–23. Access NO: 02577190 ProQuest Periodical Abstracts-Research I.
4. Kowalski, Kathiann M. "Saving Our Beaches." *Current Health 2* May 1996: 18–19. Access NO: 02789612 ProQuest Periodical Abstracts-Research I.
5. "Salvaging our Seashores." *Environment 7* Sep. 1995: 32. Access NO: 02480997 ProQuest Periodical Abstracts-Research I.

I plan several focused searches on the Internet and hope to find the most current data through this approach. Finally, I will try to find relevant pamphlets and will check SIRS to see if I can obtain literature pertaining to social implications of ocean and sea pollution.

Timetable
I intend to complete all research by August 1. From August 1 until August 5, I hope to work on rough drafts of my report—August 6 and 7 will be used to polish my work. My completed research paper will be presented to you on August 8.

Request for Approval
I request your permission to write my report on the devastating effects of ocean and sea pollution. I would appreciate any input that would benefit my research. I look forward to discussing this topic with you.

Analysis of Text Prospectus.   Figure 6.12 is a "deck" proposal—I wanted my publisher to "buy" my idea for a book. Again, it uses the memo shell, though substituting another title for "Memo." I tried to pack as much information into the subject line as possible, but did not want to exceed one line's space for aesthetic reasons. I labeled my introductory summary "Rationale," so that Christie and my other readers would expect justification for my idea. The three-concept introduction establishes that there is a problem, explains what might fix the problem, and offers my idea as the solution. The "Contents" section is an extended outline of the proposed book, and thus parallels Angela's "Areas to Be Studied" section in Figure 6.11. Following the outline of contents, a discussion of the teaching approach that the book will use explains the reasoning behind the organization of material. A summary of authorial qualifications rounds out the support section, providing further justification for considering my project. And the proposal integrates an estimate of the project's completion date with a closing sentence inviting my audience to reply. It just seemed that this closure read better as a combined statement than as two separate sections. (This brings up an important point: you want to read your writing aloud as you write and rewrite. Get used to its rhythm and you'll know when something isn't working).

Figure 6.12   Proposal in Memo-Shell Format

**Text Prospectus:
Technical Writing Handbook**

To: Christie Catalano, Sales Representative, Prentice Hall
From: Brian R. Holloway
Date: May 22, 1996
Re: Proposed Handbook for College Technical Writing Courses

Rationale
Many career-motivated students must take technical writing courses if they plan to enter health, business, engineering, science, or the social services. This very diversity of audience causes problems which characterize such courses, as a text and approach which favor one field may alienate students in another. This problem compounds when the nature of technical writing is poorly understood by some instructors; sixteen weeks of grammar drill, or letter-writing, odious in themselves, will not prepare a student to write analytical reports integrating complex material.

In order to provide maximum flexibility in the classroom, and to encourage the understanding of business communication, teachers need a short, adaptable text which

- **is not biased toward one field**
- **surveys many kinds of technical writing**
- **offers practice in revision of real writing examples**
- **builds skills cumulatively**

Figure 6.12 *(continued)*

2

Topic and Approach
Such a text must be part workbook and part "generic" treatment of the subject. **At the beginning of each section appear actual examples which reveal problems.** Exercises then ask students to **rewrite** these documents. Then the text discusses **techniques** which work.

Contents

How Does Business Writing Communicate?
- Style
- Audience
- Tone
- Structure
- Format
- Using Computers to Your Advantage

Basic Forms: Letters, Memos, Transmittals and Their Protocols
- Positive
- Negative
- Neutral

Our House to Yours
- Bulletins
- Descriptive Leaflets and Flyers
- The PSA

Basic Business Communication
- Posted Directions
- Instructional Pamphlets
- Complex Instructions and "Clustering"

Communication with Many Parts
- Writing Analytically
- Ensuring Continuity

Persuasive Communication
- Writing to Convince
- Effective Strategies for Short, Persuasive Memos and Letters

Selling Yourself
- Resumes
- Cover Letters
- Follow-up Correspondence

The Report with Many Sections
- Research Techniques
- Using Traditional and Electronic Sources
- Documentation Overview
- Writing the Survey of Background Material

Figure 6.12 *(continued)*

- Preparing Informal Evaluations and Reporting on Work-in-Progress
- The Formal Proposal
- Constructing the Complex Report
  - House Format
  - Strategies of Support and Integration
  - Working on Collaborative Documents
  - Producing a Unified Package

Estimated Length of Text: About 200 pages.

Educational Approach

The assignments in the proposed text build in complexity, chapter by chapter; also, each assignment in the research report section relies upon the last one, in "real world" fashion; the *Overview* contains material which can be used in the *Survey,* the *Proposal* relies on the previous reports for content, and the *Final Report* includes material imported from the other assignments. Such an incremental, cumulative approach assists students who use computers in their writing, as material saved on disk which constitutes a previous assignment can be retrieved, modified, and transferred to the new document. Should the final report be collaborative, students can integrate their reports on disk as well. This collaborative approach might encourage students in similar fields to work together to create unified projects.

My approach, then, affirms the future value of the work the student has just completed, and is a pragmatic one—real examples and models demonstrate what should be done. This method derives from my twenty-year teaching span and my twelve years in business.

Personal Qualifications

I helped create my first multi-part technical document in 1973 and 1974, when, as an undergraduate, I joined a committee creating a text and revamping a course in the College of Education, University of Missouri. I earned a BA and an MA in English from the University of Missouri, and a PhD in English Literature from the University of Illinois–Urbana, teaching sections of Rhetoric for Engineers, Rhetoric for Advanced Students, and Advanced Expository Writing. At Parkland College, I taught business-oriented writing, along with a full slate of other composition and literature classes. I also became a director of a retail business in 1981, handling vendor accounts, advertising, sales, and promotion, and resigning that position in 1993. I am now Associate Professor of English at The College of West Virginia, where I teach, among a diverse list of courses, Technical Writing—to positive student response. I have published several articles and received several awards. A copy of my vita is enclosed.

Development Time

I estimate a one-year development time to create the text, integrate the large amount of classroom materials, and acquire permissions for necessary items, including exemplary student work. I look forward to your reply.

Discussion of Grant Proposal. Figure 6.13 was not written to stand by itself—rather, the granting agency required a large packet of supporting material to accompany this proposal: worksheets and an extended résumé called a vita (we will discuss résumés in Chapter 8). Note that this prospectus uses a

Figure 6.13 Grant Proposal in Memo-Shell Format

PROPOSAL

TO: The West Virginia Humanities Council
FROM: Dr. Brian R. Holloway, Associate Professor of English
        The College of West Virginia
DATE: December 30, 1995
RE: Proposal for Fellowship in the Humanities

Summary

I request a Fellowship award of $2000 to assist research that would produce original scholarship and enhance the courses I teach. My request is consonant with the current *NEH Initiative on American Pluralism and Identity,* as my project overlaps literature and the study of interacting cultures. I propose to research the work of John Neihardt and Nicholas Black Elk in creating *Black Elk Speaks.* This book has exerted worldwide, multicultural influence while dramatically affecting Native American concerns and studies. Yet though it is frequently taught—and quoted—its context and origin remain unclear, the subject of debate.

Proposal: *Intermingled Cultures—*
John Neihardt and Nicholas Black Elk

I shall use the 1996 stipend to travel to the University of Missouri–Columbia to work with the John Neihardt collections at the Western Historical Manuscripts division of the Elmer Ellis Library for four weeks during the summer of 1996. I will research primary materials and artifacts produced by Neihardt's collaboration with Black Elk—a project which resulted in *Black Elk Speaks.* By examining the firsthand evidence and its context, I intend to determine the extent to which the book was a collaboration of two cultures or a compromise between them.

I have already begun this project by contacting the librarian in charge of the Western Historical Manuscript Collection at the Elmer Ellis Library. This collection houses Neihardt's original manuscripts, field notes, supporting materials, and letters concerning *Black Elk Speaks.* Also on that campus is the Neihardt Collection in the English Department's headquarters in Tate Hall; it consists of the personal library of John Neihardt, which contextualizes Neihardt's work. These collections will be available throughout the summer of 1996, when I plan to see them. Permissions will have to be granted to use or quote certain restricted materials. Though limited inspection of microfilms of certain materials through interlibrary loan is possible, it is important to examine the primary evidence, not transcriptions, for insight into the construction of Neihardt's book.

Figure 6.13   *(continued)*

2

My Objectives in Studying the Firsthand Materials Are:

To provide a scholarly basis for interpreting *Black Elk Speaks* in the college classroom, so that it may be used successfully in either a World Literature, Native American Literature, or stand-alone course;

To generate articles, culminating in a book, which will so inform others in the academic community (currently there are several contradictory books which obscure, rather than assist in, interpreting this text);

To integrate the scholarship and the text itself into English 210 (World Literature II) on our campus, a course which I have developed;

To create a separate course using *Black Elk Speaks* as the primary text. I envision a course immersing students in cultural studies, *Black Elk Speaks,* and analysis of the current influence of this work. (The "messianic" aspects of the Ghost Dance itself have a lot to tell us about long-unstable social conditions, just as Black Elk's philosophy exerts strong cross-cultural influence);

To publish the resulting course on the Internet so that other colleges will benefit from this project.

Consistency

This project derives from long association with secondary material. My involvement began in 1974 when, for high-school students, I taught Neihardt's *When the Tree Flowered* and *Black Elk Speaks.* I have used these books in advanced rhetoric courses at the University of Illinois, and have presented them in modern literature and lifelong learning classes at Parkland College. Here at The College of West Virginia, I incorporate *Black Elk Speaks* into the new offering of English 210 (World Literature and Cultures II) which I recently designed. In addition, a paper on "World Literature and the Canon" which I delivered at the fall 1994 WVACET conference discussed the book Neihardt and Black Elk wrote; this book emerged as a topic in my presentation at the Lilly Conference on Excellence in College Teaching November 17, at Miami University, Oxford, Ohio: "Enhancing the Options—Cross-Cultural Studies in the Literature Classroom." That discussion has been submitted to the *Journal on Excellence in College Teaching.*

I now wish to use the primary materials to further my study of Neihardt and Black Elk, and to benefit The College of West Virginia. I have enclosed a brief bibliography, a budget estimate, and an abbreviated resume for your consideration. I welcome your support.

Figure 6.13   *(continued)*

3

Bibliography

Black Elk, Nicholas, *The Sacred Pipe: Black Elk's Account of the Seven Rites of the Oglala Sioux.* Ed. Joseph Brown. Norman: U of Oklahoma P, 1953.

—. *The Sixth Grandfather.* Ed. Raymond DeMallie. Lincoln: U of Nebraska P, 1984.

Castro, Michael. "John G. Neihardt." *Interpreting the Indian: Twentieth Century Poets and the Native American.* Albuquerque: U of New Mexico P, 1983.

Deloria, Jr., Vine, ed. *A Sender of Words: Essays in Memorial of John G. Neihardt.* Salt Lake City: Howe, 1984.

Erdoes, Richard. "My Travels with Medicine Man John Lame Deer." *Smithsonian* 1973: 34.

—. *Crying for a Dream.* Santa Fe: Bear, 1990.

McCluskey, Sally. "Black Elk Speaks: And So Does John Neihardt." *Western American Literature* 6 (1972): 231–242.

Rice, Julian. *Lakota Storytelling: Black Elk, Ella Deloria, and Frank Fools Crow.* New York: Peter Lang, 1989.

—. *Black Elk's Story.* Albuquerque: U of New Mexico P, 1991.

Sayre, Robert. "Vision and Experience in *Black Elk Speaks.*" *College English* 32 (1971): 509–535.

Budget

Transportation	(2 x 660 miles @ .26/mile)	343.20
Meals	($10/diem)	280.00
Lodging	($145/week)	580.00
Permissions	(Estimated 16 @ $50)	800.00
Book purchases to date		170.58
Research/Writing—min. 40 hrs./week @ $25/hr.		4000.00
Total expenses, estimated		6173.78
Offset, Fellowship		<2000.00>
Contribution in-kind		4173.78

modified memo shell but deviates slightly from the familiar proposal format. Indented first sentences of paragraphs and centered headings provide a balanced, symmetrical look. Bold headings signal the beginning of key sections, and the contrast between normal and italicized type in the second heading emphasizes the concept of the project. The support section

Figure 6.14   Portion of Prospectus by Rachel Lanier

Proposal: Project in Cherokee Pottery

Timeless, delicate, and beautiful. These are just a few of the words that could be used to describe the pottery created by the Cherokee people. For thousands of years, this form of art has been practiced by the Cherokee. And like the Cherokee, their pottery has undergone many changes throughout time. Despite these alterations, however, Cherokee pottery has remained a much treasured and practiced traditional art. It continues to tell the history and stories of the Cherokee people, forever preserving a link to the past and connecting past with present and future. My work in the cultural studies program will trace the changes in Cherokee pottery throughout history and examine why it changed.

Project Focus

Cherokee pottery, like other forms of Cherokee art, has evolved in many ways. The history of the pottery is a complex and varied one. As the focus of a degree project, this history will be broken down into four sections. The first will examine Cherokee pottery before colonization manifested serious effects on the Cherokee way of life. Pottery made during this time period was undoubtedly different because of the abundance of needed materials, less reliance on the settlers for goods, and less pressure on its makers to become "civilized."

The second and third sections discuss transition. The second phase will focus on how colonization affected the creation of Cherokee pottery. Because of the onslaught of settlers and the pressure to conform to the "civilized" lifestyle, the original practice of creating Cherokee pottery was lost. The third section will analyze how the Cherokee redeveloped their art of creating pottery through assistance from another tribe, the Catawbas. With their help, and in spite of the eradication of their art form, the Cherokee began to craft pottery once again.

The fourth and final section will focus on Cherokee pottery today. For many reasons, contemporary Cherokee pottery is different from that made thousands of years ago. This segment will explore how availability of materials, tourism, and mass marketing have all affected the way in which the pottery is fashioned and used. It will also examine the implications of these effects on the future of this traditional form of Cherokee art evolving under constant pressure from the dominant culture.

Project Design

The design of this project relies on directed independent studies and readings. Through them, I will collect the knowledge and research needed to underpin a thesis. These directed studies will focus on the Cherokee pottery created during each of the four time periods. Research on how the pottery is made, different designs employed, figures on the pottery, and the intended function of each kind of piece will supplement these studies. Another directed reading will focus on the general history of the Cherokee. These committee-directed projects will provide a good understanding of how the pottery changed and why.

Figure 6.14 *(continued)*

> To enhance the design of this program, additional elements should be added. Working with Cherokee pottery presents the opportunity to apply for grants for on-site research at the Cherokee reservation in eastern North Carolina. Grants from institutions such as the National Endowment for the Humanities could be used to cover the minimal costs associated with investigation of this type. The main goal of this on-site research would be to support the fourth section's analysis of Cherokee pottery today. These grants would allow extensive research to be conducted concerning how changes such as tourism and mass marketing have affected all aspects of the art form.

(Rachel then explains her qualifications in detail, covering her general coursework and her extensive grounding in Native American culture and literature. She closes with her curriculum vita.)

summarizes the proposed project, states objectives (in an outline form similar to Angela's in Figure 6.11), and explains qualifications (the "Consistency" material). A closing paragraph follows, requesting support for the endeavor. Because the bibliography and budget were required at the end of the prospectus, I wanted to insert the closing request before readers could become distracted by those items.

Discussion of Student Proposal.   Figure 6.14 illustrates part of an academic proposal by a student, Rachel Lanier, who is explaining her intended program of study. Notice that this document is written as a short report introducing the project, detailing the project's agenda, and explaining its design. The introduction informs the reader about the subject and declares the student's purpose in studying it. The "focus" section divides the project into four phases, explaining each. The "design" element discusses courses, assignments, presentations and grant-seeking that the project will incorporate. Other sections of the report, not reproduced here, describe the student's qualifications and background. These sections serve as conclusion and reference elements. Rachel has presented her goals and pathway clearly, using the techniques of proposal and report writing.

Looking at these semi-formal reports provides a sense of the variety that can be contained in a very straightforward, simple organizational pattern. Uncomplicated organization usually works the best—even in long reports with many sections, which we'll study in Chapter 7. There, we'll learn how to develop a large report from a shorter one, how to find and integrate source material, and how to blend the different types of transactional writing to form a seamless unity.

Assignments ===================================

1. Your mission is to visit a local business—a chain store, mall store, or independent concern—and take notes about a procedure you observe that you think needs improvement.

   Next, develop these notes into a rough-draft summary of the problem to be fixed. Try to think analytically (what are the features of the problem?) rather than narratively (first I saw this, then I saw that). Creating an outline after using an idea web will help you think in terms of topics, not events.

   From that summary, construct a one- to two-page informative report in a letter or memo-shell format. Use headings in the support section to divide the problem into its components. Be sure to develop with detail your description of the problem's different facets.

   Your report should be addressed to a specific audience for the best focus: perhaps a customer service representative. You yourself will need to assume a role—consultant, district manager, vice president, perturbed client.

   Remember that your report should have a message section stating that a problem exists. The support section should summarize the problem feature-by-feature. The closing section needs to recommend that the problem be fixed.

2. Recast the descriptive report you wrote in Problem 1 into a problem-solution report. Write it as a proposal, using a memo-shell format. Your message section—called "Summary"—should explain the *purpose* of your document and state the new policy you wish to enact to prevent future difficulties. The support section must *outline the features* of the problem, discussing them analytically. Next, *explain the benefit* of following your suggested procedures, using comparable examples pertaining to other businesses as references. Discuss the amount of *time* (and cost) estimated to correct the difficulty, and *close* with a paragraph recommending that your reader both follow the proposed plan and reply to your communication.

3. Present the reports you created in Problems 1 and 2 in small groups—make sure everyone reads all the reports in the group and comments on them. Assess how well you analyzed the problems and how effectively you proposed solutions.

4. Write in letter or memo form a one-page report explaining to the class what social, technical, or business issue you intend to research for your long report (coming up in Chapter 7). Focus your report as specifically as possible by narrowing the problem down. For example, don't write about global deforestation, but about deforestation in the Pacific Northwest. Explain to your audience the features of the problem you wish to research; what do you think are its components?

5. Develop the document you created in Problem 4 after you've browsed in the library. Expand the document's message, support, and closing sections into a prospectus similar to that shown in Figure 6.11. Then present this document to the class for discussion in small groups.

6. Write a one-page report that explains to other students *how to research* your specific topic. This report must be in a form approved by the class so that each person's report is identical in format. The reports will be combined with others to create a resource guidebook for students who want to know the best way to research different topics. Typically, such reports divide under three headings:
   a. Explain the topic.
   b. Explain the best method of research and discuss pitfalls to avoid. What search media generate the best results? Which libraries were useful?
   c. List specific articles, books, abstracts, and journals that you find helpful.

7. As a class, determine appropriate groupings for your reports. Classify the reports under headings that are subtitles of sections in the guidebook created in Problem 6: Business, Social Issues, and other general terms. Construct a table of contents, add a cover sheet, and have the document copied for distribution on campus.

CHAPTER 7

Writing the Formal Report

Features of the Formal Report

Overview

Semi-formal reports, as we have seen, develop from the memo format and incorporate bold headings, bulleted lists, and straightforward patterns of organization. What differentiates such reports from formal ones?

Formal reports also depend on the memo-shell format for internal design, but integrate most of the other types of writing that we have studied so far. Traditionally, the formal report consists of these components, usually in the following order:

1. A binder holding the report together
2. A cover sheet prepared for maximum visual impact
3. A letter or memo, called a transmittal, which addresses the audience and explains the report
4. A table of contents (followed by a list of illustrations, if important)
5. An abstract or executive summary providing the reader with a condensed overview of the contents
6. The report itself, containing an introduction, support sections, and a closing section
7. Any attachments or appendices to the report

If such a report is sent as an e-mail attachment, the e-mail often takes the place of the covering letter or memo and serves as the transmittal. Also, though a binder is obviously not used in such cases, formal reports sent electronically often use elaborately designed images as "cover sheets."

Binding

Formal reports are often attractively bound. This enhances visual appeal and ensures durability as long as the stiffer plastic binders are used. Cheap, flimsy binders with slide-on spines send the wrong message to your audience. You don't want your report coming apart in your client's hands, and you don't want your client to feel deemed unworthy of the few cents a better cover would cost. Plastic comb-binding works well for thicker reports, but tends to distort the shape of documents of fewer than thirty pages, making such texts look wavy. This may not be the right message to send to your client, either.

An alternative for those reports is a proprietary process called Velobinding, in which two thin, strong plastic strips secure the pages of the report. The strips interlock through holes punched with a special tool along the left edge of the document. To make the report look even more classy, take two acetate sheets used to make projection transparencies, place one over the cover page of the report, put one after the last page of the report, and then bind the whole together so that the text and the graphic on the cover page show through. For an interesting visual effect, use a colored set of transparencies. The cover sheet's graphics and title, revealed through the plastic, will gain intensity. Of course, the protocol of your profession and institution will determine the choice of binder.

If the binder—of whatever kind—is opaque, the folder must be labeled with the title, the words "Prepared for" followed by the recipient's name, the words "Prepared by" followed by your name or your firm's name, and the date. Use taste in choosing a label—or better yet, make one on your computer, giving it a border and using a type and font identical to that of the cover page. Figures 7.1 and 7.2 illustrate types of binders and labeling, respectively.

Figure 7.1   Illustrations
of Binders

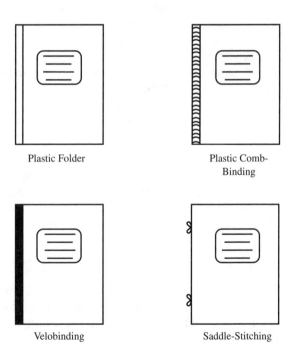

Plastic Folder

Plastic Comb-
Binding

Velobinding

Saddle-Stitching

Figure 7.2 A Typical Label

> **Ground Fertility Recovery**
>
> Prepared for: John Schmendrick
>
> Prepared by: Newman Contracting, Inc.
>
> Date: June 23, 2004

From the start, then, you want your report to achieve maximum impact. Close attention to the "package" itself ensures that. Even if past practice has been to send out formal reports stuck together with giant staples, avoid doing this.

Booklet Format. Your report might, however, be produced in booklet format, in which case it will be "saddle-stitched." Not stitching but staples are used in this process. Here's how it works. You've prepared your report using a program designed to give you control over the pagination of text and the positioning of the whole text on the page. You then create a master copy of the report—typically laying out corresponding 8$\frac{1}{2}$-by-11-inch pages two by two to produce 17-inch-long sheets. A high-quality copier duplicates the reports in unfolded 17-inch format. Before folding each report, a cover sheet produced on a color copier is added. The report's spine is stapled twice with an extra-long stapler, and the resulting 8$\frac{1}{2}$-by-11-inch booklet is creased down its new spine. Ragged right-hand edges can be trimmed on a paper-cutter for a clean look (this is why you need a program that controls the drift of the text on the page, or you would be cutting off text during this process). Such booklets are ideal for distributing at small conferences (Figure 7.3).

Whichever method you use to produce the report package, however, the result should be professional in all respects. Immediate visual appeal provides a sense of quality and attention to the client; slovenly appearance does not.

Cover Sheet

The cover sheet is the first item inside the bound package. Centered on it appear:

The Title of the Report
Prepared for: [the Recipient]
Prepared by: [You or Your Organization]
Date

Also, the cover sheet may contain an appropriate illustration to get the attention of the recipient. Illustrations that have nothing to do with the subject matter of the report are useless, and in fact may deflect your client's attention. We'll examine some effective cover sheets in the upcoming gallery of model reports. Beware of overworked clip art, though. Figure 7.4 illustrates a sample cover sheet.

Figure 7.3 Making a
Booklet in Your Office

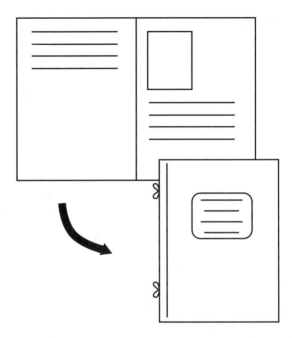

Seventeen-inch masters with text laid out in correct sequence
are duplicated on the copier, collated, and attached with two
staples on the spine.

Transmittal Letter or Memo

If the report is being sent physically, and not as an attachment to an e-mail that
serves as a transmittal, you will have to include a transmittal letter or memo.
Many firms place this document first—attaching it to the outside of the report.
There is nothing wrong with that; however, in the process of opening mail, such
items can get thrown out with the envelope. I recommend including the transmittal letter or memo in the report, right after the cover sheet. Remember that
if the report is being sent externally, you should use a transmittal letter; if the
report is an internal document, a memo would normally be used.

What should the letter or memo contain? The transmittal should:

1. Address the recipient by title and surname.
2. Provide a brief summary of the background leading up to the report.
3. Mention the key findings of the report.
4. Explain the benefit to the reader provided by the knowledge contained
   in the report.
5. Indicate those not mentioned at the beginning of the letter or memo to
   whom copies of the report have been sent: a line at the bottom does
   this—just write c: (for copy) followed by names of other recipients. (You
   may still see cc used for this purpose, even though cc technically means

Figure 7.4 A Sample Cover Sheet

Insurance Options

Prepared for: Kathy Hough

Prepared by: Bayard Siscomb

Date: March 4, 2004

"carbon copy.") The purpose of doing this is ethical; people have the right to know about other recipients of apparently confidential information.

Note: If you are using a simplified letter as the transmittal, you should—as you would do with a memo—state the title of the report as your subject line. For reference purposes, consider the examples in the Gallery of Documents beginning on page 143.

Table of Contents

This section puts *parallelism* to work—you don't want to mix nouns with verbs since that would deflect the reader's attention. All elements in the table of contents constitute an outline of the report using grammatically similar phrases.

You want to watch the amount of *detail* in your table of contents, since you need to provide readers with a true picture of what your report actually emphasizes. Absence of detail in key sections and a proliferation of detail in less-significant sections will misrepresent the scope of the report.

You'll need to assign *page numbers* to the items in the table of contents so that the reader may find everything quickly.

Table of Illustrations

If there are only a couple of illustrations, the best plan is to include them in a separate list—below the table of contents, but not on a new sheet. Typically you'll need the illustration number (such as Figure 1 or Table 4), the illustration caption, and its pagination. But if there are many illustrations, you'll need a separate page for such a table. In fact, you may have many illustrations of several types, necessitating a table that classifies them as:

◆ Figures (drawings or drawings with text; Figure 7.5)
◆ Tables (tabular arrangements of data; Figure 7.6)
◆ Plates (photographs; Figure 7.7)

Whether you need anything as elaborate as this classification depends on whether you have a report that emphasizes visual material.

Abstract, or Executive Summary

This *must* be a summary, *normally* not to exceed 250 words. You don't want to open up a large report only to find a summary that is itself a huge document! The executive summary must communicate to the topmost reader of your report the essence of that report. Employ analytical writing. Use a message-support-closure strategy, not a narrative one, and try to break up the supporting material into a bulleted list for easy digestion.

Figure 7.5   A Figure

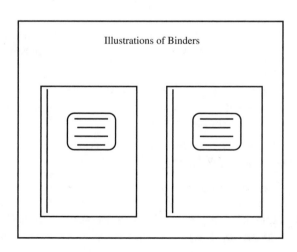

Illustrations of Binders

Figure 7.6 A Table

Table of Gauges— Custom Metal Thicknesses		
String	**Diameter**	**Wound/Plain**
12	.026	Wound
11	.048	Wound
10	.020	Wound
09	.036	Wound
08	.014	Plain
07	.030	Wound
06	.009	Plain
05	.020	Wound
04	.014	Plain
03	.014	Plain
02	.011	Plain
01	.011	Plain
Diameters in decimal fractions of one inch.		

Figure 7.7 A Plate

Proserpine.

A "plate" may be a reproduction of a photograph or a facsimile of a work of art. In this case, the plate reproduces an engraving from a nineteenth-century book. *Source:* Bulfinch, Thomas. *The Age of Fable or Beauties of Mythology.* Ed. J. Loughran Scott. Philadelphia: McKay, 1898, page 69.

Some business writing teachers would qualify the above statements and suggest that if the executive summary runs long, a separate abstract or summary should precede it. I would caution against doing this routinely; it provides an abstract to summarize a summary that summarizes a report (which itself has an introduction containing summary elements). Do not inflict excessive preliminary material upon busy readers. Work on your executive summary to make sure it reflects the gist of the report within 250 words.

Many inflated executive summaries sprawl beyond the 250-word limit because they:

◆ Include either patronizing or servile diction that should be eliminated (such as "As you will understand if you read Section Three of the Report . . ." or "Our hard work on this project, especially shown in Part Two . . .").
◆ Cannot distinguish between main and subordinate ideas, and thus give both equal space in the summary.
◆ Were written before the rest of the report was complete and thus lack a clear sense of the agenda of the completed report. Write your letter/ memo of transmittal and your executive summary at the same time: *after* the report has been written and revised to assume its final form (after you know exactly where you're going, because you've been there).

Examine the models in the Gallery of Documents (page 143) for guidance concerning executive summaries.

Introduction

Remember the three-concept introduction we looked at in the previous chapter? We noted that the beginning of a semi-formal proposal might explain:

◆ That there *is* a problem (background)
◆ What might *fix* the problem (description of potential remedies)
◆ The specific *remedy* to be proposed (the thesis or purpose of the proposal)

People like to think in threes: remember all the folktales in which things recur in multiples of three—or the basic structure of transactional writing, with its three-part system of message, support, and closure. A tripartite introduction to a formal report may be both psychologically satisfying and memorable. For this reason, it is commonly, even instinctively, employed. Consider these patterns:

Introduction to an Informative Report

◆ *Statement* of purpose of the report, investigation, or activity
◆ *Description of the role* and scope of the undertaking
◆ *Explanation* of how the report addresses the different *parts* of the issue

Introduction to an Informative Report

◆ *Discussion* of the background of the report, project, investigation, or activity
◆ *Description* of the particular task, project, or investigation
◆ *Key findings* generated by the undertaking

Introduction to a Problem-Solution Report

◆ *Overview* of problem
◆ *Statement of purpose*—that the report will show how to fix the problem
◆ *Thesis* of report urging a specific solution

Introduction to a Feasibility Report

◆ *Overview* of situation
◆ *Description* of particulars designed to address the situation
◆ *Reasons* why the suggested particulars are feasible

Note the general features of these outlines of typical introductions. The introduction surveys the issue, explains the parameters of the investigation or report, and enumerates the salient components necessary to resolve or understand the concepts presented. That enumeration may be accomplished within a bulleted list, the items of which may be subheadings within the body of the report.

Support Section

An understanding of claim and substantiation is crucial to the success of your support section in communicating the ideas that reinforce your message— contained in the introduction, the abstract, and the transmittal. Generally, support in a formal report follows a structure in which assertions of fact, reinforced by values, produce statements of policy.

Assertions of fact are not the facts themselves, but general statements based on the specifics to be shown. For example, a topic sentence at the beginning of a paragraph may be such an assertion, supported by the specific facts that constitute the body of the paragraph. Or—as is common in transactional writing—the header itself may contain the assertion of fact that its following paragraph supports. Remember that technical writing commonly presents the specifics in the form of bulleted lists underneath a general assertion, *or as illustrations*. Such specifics include

◆ Statistics
◆ Case histories
◆ Definitions followed by examples
◆ Narrative or descriptive summaries
◆ Process explanations
◆ Analyses of other events
◆ Reasons derived from logical argument
◆ Charts, graphs, and tables

Values held by the audience and encouraged by the writer may help to reinforce the logic of support. For example, a report on the need for developing a youth recreational center probably assumes that both writer and audience desire good things for the community's children. Or, supported assertions of fact may suggest that the audience modify its value judgments in light of the new data presented.

You must be sure of your audience so that you can handle the reinforcement of values properly. The hardest task in transactional writing is creating a report for an unknown audience who may be entirely skeptical about or overtly hostile to your goals.

Assertions of fact—given the right audience reception—should lead to a statement of *policy* in a report with a clear goal. What do you want your audience to do? If the report's goal is the understanding of information that you present, then that is your policy. If the report's goal is to convince the audience to act, that action is your policy. Generally, transactional reports begin with sections devoted to analysis and presentation of facts; then they explain or reinforce the policy—the message.

Closing Section

Often the specific statement of policy appears in the closing section of the report. If the report is informative, then the conclusion expresses the wish that the readers understand the information presented and apply this information to answer their needs. If the report is persuasive, the conclusion requests either action or acceptance.

A list of any sources used follows the conclusion. Some institutions place the list two spaces below the text of the conclusion; others, especially for long reports, begin the list on a new page. Of course, academic reports normally follow the directions of the style manual in the appropriate field—for example, APA, MLA, or Chicago style—in positioning reference lists.

Back Matter

A formal report always has *front matter* that precedes the introduction; it may contain *back matter* as well. Back matter, if present, typically consists of "raw" documents not integrated into the report but relevant to it. For example, a college student researching the feasibility of a day-care center on her campus might accumulate the budget proposals and the minutes of an ad hoc committee that had considered developing such a facility in the past. She might discuss material from the raw documents in her report, but may also wish to include the entire documents as appendices or attachments, as their presence in the report ensures credibility. She would not repaginate these documents, but would caption each with the word "Appendix" and the number I, II, or III. In the body of her report, she could refer readers to the appropriate appendix using a parenthetical reference: (see Appendix I). Back matter does not need to be limited to such documents, either; surveys, marketing investigations, or brochures may be placed at the end of a long report depending on need. Note that standard practice in business and technical reports mandates placing appendices *after* a list of sources, not before—as they would be in an academic research paper.

Understanding the Role of the Persuasive Pattern

Authors constructing reports explaining or defending a position frequently enhance their documents by incorporating elements from classical strategies of persuasive discourse. Those writing informative reports should study this material as well, since any report can benefit from the audience-centered approach that the classical persuasive model provides. We surveyed this model briefly in Chapter 2; now, we'll apply its principles to better understand the creation of formal reports. Figure 7.8 demonstrates the relationship between classical rhetoric and modern persuasion.

Figure 7.8   Structures of Persuasion

Persuasive presentations (essays, speeches, reports) borrow their organization from classical rhetoric. In classical rhetoric, a presentation has the following parts:

> **Exordium.** The first part of the introduction is an *exordium,* or a direct acknowledgment of your audience.

> **Narratio.** Then follows an illustration of the subject, which leads to—

> **Propositio.** The thesis of the presentation. This ends the introduction.

> **Divisio.** The first part of the body defines key terms, or presents the structure of the argument to the audience ("divides" the discussion).

> **Confirmatio** and **Confutatio.** These parts of the body explore problem/solution, or pro/con. This is the heart of the argument.

> **Conclusio.** The conclusion, which should refer to the **narratio** if possible.

In modern persuasive writing, we can use these ancient divisions of presentation to control our argument:

> **Introduction.** A good introduction does three things—it
> *Addresses* the audience
> *Illustrates* the problem
> Proposes the **thesis.**

> **Body.** The body of a problem/solution paper
> Begins with a section that *defines* unfamiliar terms or shows the *division* of the argument to come
> Discusses the **problem**
> *Discusses and evaluates the solutions.*

> **Conclusion.** The conclusion calls for action or belief and—if artful—refers back to the *illustration* used in the introduction.

Well and good, you may say—but how does the classical model influence report-writing? It does so in several ways:

♦ The cover sheet and the letter or memo of transmittal act as a preliminary *exordium,* providing the first contact with your audience and getting the audience's attention immediately. This initial step, necessary for the report to be routed to the correct parties for review, ensures that the report will be perceived as written to *somebody*—not to a general, unspecified audience.

♦ The executive summary assumes the role of a preliminary *narratio,* illustrating the subject by stating its gist so that the reader will be prepared for the report to follow and be able to make general decisions based on the information.

♦ The introduction addresses your audience (*exordium*), reviews the issues (*narratio*), and declares findings or proposes an agenda (functioning as the *propositio*).

♦ The support sections combine the functions of *divisio, confirmatio,* and *confutatio.* They divide up the issues and discuss them. The report body frequently begins with a definitional section so that the reader will understand key terms used throughout the report. Or the body may commence by dividing the issue tackled by the report, breaking it down into its components so that the reader understands the structure of the report. At times, the introduction outlines the support to follow, using a bulleted list. Then the body may develop its argument—whether persuasive (in which case it most likely will resemble the classical problem-solution structure) or informative (in which instance the support will explain the content of the report).

♦ The closing section in the formal report corresponds to the classical *conclusio.* Remember that it is smart rhetorical policy to refer to the initial illustrations used in the introduction here, reminding the reader of their significance to your agenda and project.

Obviously, if the introduction, body, and conclusion of your report have a persuasive agenda, they may closely resemble the classical pattern. An informative report, however, often still requires specific locations in its structure to acknowledge an audience, employ an initial illustration, state its agenda, divide its scope, define key terms, and partition its argument.

Developing Long Reports from Shorter Ones

In practice, many of the decisions about where to place what in the report derive from the evolution of the report from a series of shorter documents. Perhaps this sequence takes place:

1. A memo or letter of inquiry initiates the project; either someone within the organization needs to know something or an outsider has contacted the organization wanting information.

2. Responding to that preliminary communication, a team prepares a short, exploratory informal report for internal use. Staff members review the document, making suggestions about content, approach, and agenda.
3. Using that document as a foundation, writers prepare a project proposal as a semi-formal report sent externally. The new document proposes that the organization undertake a project. The proposal is accepted.
4. The team creates a formal report on the project.

It could be said that the formal report has grown from the exploratory writing of earlier phases of the project. Of importance too is the fact that the later documents may borrow wording and concepts from the earlier ones (in the Gallery of Documents, note particularly how Angela Stewart's informal report shown in Figure 7.14—a memo—develops into her proposal, and how that proposal influences the construction of the final report).

Methods of Research

Strategy

So you need to locate material for your report? Will you use the library or the Internet? Actually, that choice is not a dichotomy since many libraries have an Internet presence and many libraries use databases accessible over the Internet. But library use can be awkward when one visits a facility and wanders around hoping that books, periodicals, and the other "right stuff" will leap off the shelves and into one's arms. Such unstructured browsing is unnecessary. Today's libraries offer phenomenal shortcuts for the researcher—and, of course, superlative sources, not the least of which is the librarian. Ask the librarian whenever you can't locate something. And remember the following taxonomy, or classification, of research materials.

Types of Research Materials—Retrospective and Current

Resources generally divide into two main types; retrospective and current. You need to look at the right type to find appropriate information.

Most—but not all—books are retrospective, meaning that they "look back" over a period of time, that they do not usually contain "up-to-the-minute" data, and that they are valuable for acquiring a detailed, historically based understanding of a topic. The logistics of book production reinforce this; it may take several years for a book to reach its market, and during that time much about the subject may change. Reference books such as general and specialized encyclopedias and almanacs provide historically retrospective information and occasionally—in the case of some works—fairly current material as well. Unless you are doing comparative historical research, look for the latest editions of such references.

In the old days, subject and author catalogs full of index cards helped you find the books you needed. Now, libraries have adopted electronic card catalogs. Most of these systems—such as Athena—are self-teaching and easy to use. You simply enter the author's name or the key words of the subject, and the database generates a list of appropriate sources that you can print out. Customarily, such online systems will tell you the shelf (and library branch) locations of the sources, let you know whether or not the books are checked out, and enable you to browse related topics or titles consecutive in shelving. Many libraries have comprehensive indexes online, such as the University of Missouri's Merlin, which allows you to search for books and articles using subject or author classifications. Some colleges and universities allow access to their holdings catalogs through the Internet.

Libraries do not have to be enormous to possess efficient electronic catalogs. For example, figures 7.9 and 7.10 show two stages in a search for books via an electronic database that catalogs the contents of a small library. The health sciences student, interested in shamanism and healing, first asked the catalog if there were any occurrences of the word shamanism in titles in the database; the catalog obliged by listing the occurrences. The student next called up the list of potential sources, tagged with the shelving information so that they could be easily found. A more restricted search would result from the use of both terms—shamanism and healing—since that would narrow the scope of inquiry.

Figure 7.9 First Step—A Student's Book Search Using The Library Corporation's BiblioFile Intelligent Catalog®

BiblioFile Intelligent Catalog		
Occurrences of: SHAMANISM	Type /	Works
Imagery in healing: shamanism and modern medicine/ Jeanne Achterberg.	TITLE	1
Planet medicine: from Stone Age shamanism to post-industrial healing/by Richard Grossinger.	TITLE	1
SHAMANISM	SUBJECT	4
Shamanism: archaic techniques of ecstasy. Translated from the French by Willard R. Trask.	TITLE	1
SHAMANISM — NEPAL.	SUBJECT	1
Shamanism; the beginnings of art.	TITLE	1
SHAMANISM — TIBET.	SUBJECT	1
SHAMANISM — UNITED STATES — BIOGRAPHY.	SUBJECT	1
Cover subtitle: A journey into the world of spiritual healing and shamanism.	NOTE	1
*** End of occurrences ***		

Figure 7.10 Second Step—A Student's Book List Derived from The Library Corporation's BiblioFile Intelligent Catalog®

Title: Imagery in healing: shamanism and modern medicine/Jeanne
            Achterberg.
Publisher: Boston: New Science Library, Shambhala; New York: Distributed
            in the U.S. by Random House, 1985.
Collation: viii, 253 p.: ill.; 23 cm.
Location: R726.5 .A24 1985

Title: Planet medicine: from Stone Age shamanism to post-industrial
            healing/by Richard Grossinger.
Publisher: Garden City, N.Y.: Anchor Press, c1980.
Collation: 390 p.; 21 cm.
Location: GN296 .G76 1980

Title: Shamanism; the beginnings of art.
Publisher: New York, McGraw-Hill [1966, c1967]
Collation: 175 p. illus. (part col.) 27 cm.
Location: GN477 .L613

Title: Shape shifters: shaman women in contemporary society/
            Michele Jamal.
Publisher: New York: Arkana, 1987.
Collation: xx, 204 p.: ports.; 20 cm.
Location: BL458 .J36 1987

Title: Healing states/by Alberto Villoldo and Stanley Krippner;
            foreword by Lynn V. Andrews.
Publisher: New York: Simon & Schuster, 1987.
Collation: xvi, 207 p., [16] p. of plates: ill.; 21 cm.
Location: RZ400 .V5 1987

Current materials include newspapers and magazines (for popular consumption), journals (for academic or specialized interest), and the contents of various CD-ROM and online databases or "search engines." Each has a different use and risk of use.

Newspaper and magazine articles may provide relatively superficial information unless written by specialists commissioned to provide thoroughgoing detail. Articles contemporaneous with the event described might not contain all the information that later investigation yields. But magazine and newspaper articles can be useful in assessing reactions to an event immediately following that event. They tend to be good indicators of popular sentiment, and they can alert the researcher to experts and notables in a field.

Journals provide detailed, scholarly analyses of specific academic questions. A journal's focus upon narrow but deep areas of inquiry makes it useful for understanding current hypotheses about an issue as well as gleaning appropriate detail. But because of the time lag involved in getting journal articles into print, these scholarly papers may sometimes lack absolute currency. It is also important to follow up on a subject once it appears in a journal, tracking letters and subsequent articles that react to the initial published paper.

Such periodical materials used to be indexed in hardbound general and specialized publications. Over the years, one very popular index for the general reader has been the *Reader's Guide to Periodical Literature*, which provides lists of articles under subject headings such as "Pollution" and "Education." Specialized hardbound indexes such as nursing, education, business, and criminal justice compilations have been useful for researchers in those fields, while *NewsBank* and the *New York Times Index* have ably assisted researchers looking for newspaper articles preserved on microfilm or microfiche. Generally, it helps to know the official Library of Congress subject heading for your topic (your library has these headings listed), though you should try alternatives.

Of course, the new generation of CD-ROM and online indexes of periodicals can greatly accelerate your research. When researching, choose a database that offers a full-text feature, enabling you to print out the entire article you've selected; these include not only *NewsBank* but *ProQuest, SIRS,* and *EBSCO* (formerly a CD-ROM database, but now online through the Internet). Browse the topic using buzzwords, trying different phrases to make sure that you cover all the possibilities. For example, "Native American art" may locate some entries in the database; "American Indian art" may get others. Once the database has provided you with a list of possible articles, read the abstracts of articles provided. Select the articles you wish to use, and print those for your reference.

Figures 7.11, 7.12, and 7.13 trace the path of a student's search for information about hypertext. First, the database builds a bibliography of promising articles in its inventory. Next, the student considers summaries of interesting articles provided by the database. Finally, the student selects the full text of one such article and prints it for reference.

The Internet itself is a superb source of good, bad, and indifferent current information. The format of general searching on the Internet is similar to that of investigating a topic using a fixed database such as one of those described above: one selects a "search engine" from those offered, types in the subject word or words, and awaits the production of a list of materials or sites in the contents of which the key word or words occur. Some search engines, such as Lycos, provide abstracts of the articles and postings found. Google, HotBot, and Yahoo! are among the helpful search tools. A good starting point in any Internet search is the Internet Public Library (at **http://www.ipl.org** or just type **Internet Public Library** in the query box of a search engine). The Internet Public Library has reference sections and links to specialized indexes.

Figure 7.11 Using a Database—Building a Bibliography,
Keyword = Hypertext

ACADEMIC ABSTRACTS FULL TEXT   Bibliography   Page 1

1. LITERACY
   Technology and learning.
   (Publishers Weekly, 12/13/91, Vol. 238 Issue 54, p20, 2p, 1 bw) (0000-0019)
   (The library has this journal)

3. WINDOWS (Computer software)
   Windows development tool kit.
   (PC Computing, Nov91, Vol. 4 Issue 11, p290, 10p, 1 illustration, 5 c)
   (0899-1847)
   (The library has this journal)

5. SMARTEXT (Computer software)
   Lotus SmarText lets you build fast, easy **HYPERTEXT** *applications.*
   (PC Computing, Sep91, Vol. 4 Issue 9, p80, 2 p, 2 c) (0899-1847)
   (The library has this journal)

6. MAPPING **HYPERTEXT** (Book)
   Mapping **HYPERTEXT**.
   (Whole Earth Review, Summer91 Issue 71, p59, 1/2p, 4 diagrams, 1 bw)
   (0749-5056)

12. **HYPERTEXT** (Computer operating system)
    The smart tool for information overload.
    (Technology Review, Nov/Dec90, Vol. 93 Issue 8, p42, 8p, 1 illustration, 1
    c, 8 bw) (0040-1692)
    (The library has this journal)

14. TEACHING
    Reading and technology in the future.
    (Reading Teacher, Nov90, Vol. 44 Issue 3, p262, 2p) (0034-0561)
    (The library has this journal)

15. GUIDE (Computer software)
    A worthy guide to **HYPERTEXT**.
    (Personal Computing, Aug90, Vol 14 Issue 8, p129, 1p, 1 chart, 1 bw)
    (0192-5490)
    (The library has this journal)

19. READING—Study & teaching
    HYPERTEXT *and hypermedia: Discovering and creating meaningful learn-*
    ing environments.
    (Reading Teacher, May90, Vol. 43 Issue 9, p656, 6p, 4 bw) (0034-0561)
    (The library has this journal)

32. XANADU (Electronic publishing network)
    (Omni, Sep87, p16, 2p) (0149-8711)
    (The library has this journal)

Source: EBSCO Publishing, Ipswich, Massachusetts

Figure 7.12 Example of a Source Summary Generated by the Abstract Function in EBSCO

Subject:	ENGLISH language—Study & *teaching*
Title:	Reading hypertext: Order and coherence in a new medium.
Author:	Slatin, J.M.
Summary:	Argues that hypertext is very different from more traditional forms of text, that the differences are a function of technology and are so various as to make hypertext a new medium for thought and expression. Taking the *computer* actively into account as a medium for *composition* and thought; Predictability; Negotiating the three types of hypertext readers; 'Non-sequential writing' in an open dynamic system; Linking enormous quantities of material.
Source:	(College English, Dec90, Vol. 52 Issue 8, p870, 14p)
ISSN:	0010-0994
Item No:	9012241371

** The library does not have this journal **

Source: EBSCO Publishing, Ipswich, Massachusetts

Figure 7.13 Pathways Through Electronic Research

Type	
Database or Index	**Search Engine**
Select from a pool of information to which the index subscribes.	Choose the probable sites using a key term.
Steps	
Insert subject words in query box.	Insert subject words in query box.
View list of available articles.	View groups of likely sites.
Evaluate abstracts of such articles.	Select promising sites.
Read desired articles.	Evaluate site material.
Print copy or download to disk.	Print copy or download to disk.
Examples	
Use the word "hypertext."	Use the word "hypertext."
Find list of articles about "hypertext."	Find list of sites using "hypertext"—some search engines provide abstracts.
Study abstracts of articles in which "hypertext" occurs.	Study promising sites using "hypertext."
Read desired articles.	Bookmark sites so you can return to them—use the "bookmark" function.
Print/download needed articles.	Print/download needed articles.

Do be aware that anyone, no matter how depraved or silly, may have a site on the Net; however, know that the Internet is also a first-rate source for absolutely current information—including corporate postings, job openings, market statistics, CNN News, and travel and weather information. Retrospective texts abound on the Net as well. Just surf carefully!

Primary and Secondary Materials

Just as the distinction between retrospective and current sources is important in defining the scope of your research, so is the classification of sources into primary and secondary items.

◆ Primary sources are firsthand or as close to the event as possible. For example, a team that conducts an experiment writes an article about it in a scholarly journal. That source is a primary one.

◆ Secondary sources report on primary sources, usually summarizing them. An article in a magazine discussing the journal article mentioned above is a secondary source.

◆ Generally, if both primary and secondary sources are available, primary ones are trusted more for their validity. This doesn't negate the worth of good secondary sources, which may analyze primary ones and contribute their own input to an ongoing discussion. But some secondary materials clearly dilute the content of the primary sources they discuss. Popularized, watered-down secondary versions are less trustworthy. You owe yourself and your reader an examination of the primary materials connected with your topic.

◆ A source summarizing several secondary sources might be valuable for research leads or its bibliography, but does not carry the authority of materials closer to the original event.

Handling Sources

Overview

This section reviews the use of outside sources. You should be familiar with the principles of reference format from your English Composition course; however, if you are not, consult a style handbook such as the *Prentice Hall Reference Guide to Grammar and Usage,* by Muriel Harris. Or, for MLA practice, refer to the current edition of Joseph Gibaldi's *MLA Handbook for Writers of Research Papers.* APA format appears in the latest *Publication Manual of the American Psychological Association,* and a clear overview of the University of Chicago style of documentation, as well as much other helpful material, is located in *A Manual for Writers of Term Papers, Theses, and Dissertations,* by Kate L. Turabian (with revision by John Grossman and Alice Bennett). Depite the different formats employed, the philosophies and goals of such documentation systems are similar.

Documentation Systems—Citation and Reference List

You use the same documentation system throughout the report, just as you did in writing a research paper for English. If the report marks borrowed ideas with MLA parenthetical citation, the report's ending list of references should also be an MLA Works Cited page. If APA citation is used, the reference list, in APA format, should be entitled References. Current MLA and APA practices avoid footnotes, except for explanatory material that does not fit into the text of the paper itself but is still necessary for the reader to know. Sometimes you will see a "house style" used by organizations in which the sources listed at the end are numbered as well as alphabetized; the borrowed material in the text of the report is followed by the appropriate number keying the material to the appropriate source rather than by a parenthetical citation. Some organizations—though otherwise using MLA or APA format—single-space the text of each source entry in the bibliography and double-space between entries. Follow the requirements of your institution (or course) for consistency; note examples in the Gallery of Documents at the end of this chapter.

What Should Get Cited?

Remember, anything not generally known to your audience, or in the realm of public knowledge, needs to be tagged with a citation—whether that is an MLA parenthetical citation of author and page number, an APA citation of author and date, or a reference number. In other words, such an idea *belongs to somebody* who needs to be credited for it—whether you are quoting verbatim, paraphrasing, or summarizing to reexplain the concept.

Integrating References

Remember too that this piece of information must be integrated into your own writing. For example, be sure to introduce a quote with a transition; a quote that stands by itself as a complete sentence looks as though it has been glued or taped onto your text.

Not this:

"Allen measured the amount of acid rain in the Northeast" (Jones 43). [This is a "dumped quote" lacking transitional connection to the text.]

Instead, this:

One scientist, Dr. John Allen, "measured the amount of acid rain in the Northeast" (Jones 43). [An explanatory transition relates the quoted material to the text of the report.]

Exercise

1. Research a likely topic for a longer report by browsing through listings of traditional and electronic sources, seeing what information is available. Write a short report of 250–300

words explaining to the class whether there is enough material to research your topic, and what sources look promising. Use memo format (refer to Figure 7.14 for an example).

2. Critique these reports in small groups in class; suggest avenues for research and ways of focusing your topic. After noting the comments of your classmates and performing further research, write a two-page research proposal explaining:
   - What kind of report you will write (i.e., will it be informative or will it discuss a problem and offer a solution?)
   - What the specific topic of the report will be
   - What the subtopics of your report will be
   - What sources you think you will use
   - In what time frame you will complete the assignment

   Refer to Figure 7.15 for formatting ideas.

3. After the class has reviewed the proposals in small groups, research and write a formal report of four to seven pages on your topic. Study Figure 7.16 and others in the Gallery of Documents for suggestions about format.

Gallery of Documents

Ending this chapter are a number of documents that help to provide information about creating the long report and its supporting materials.

Figures 7.14, 7.15, and 7.16 show a preliminary report, research proposal, and short formal report, respectively, written by Angela Stewart. Note these aspects:

◆ Elements of the preliminary report—which contains the agenda and some bibliographic material—are reintroduced and reapplied in the proposal (or "Prospectus").

◆ The proposal then sets forth the sections to be developed in the future paper and presents a working bibliography.

◆ The final report uses the proposal's wording in its front matter and expands upon the points presented in the proposal, which now become main headings of sections.

◆ Bold type and the use of extended lists make the final report reader-friendly.

◆ The final report "grows" from the initial report.

Figures 7.17, 7.18, 7.19, 7.20, and 7.21 contain sections of a draft of a formal report by Douglas M. Burns, a microbiologist. This document summarizes existing research and proposes a new project. The complete report's many parts derive from technical format, though driven by the protocol of work in a specialized field. The main units of this multi-sectioned report include:

◆ A **cover sheet** acting as both transmittal describing the purpose of the document and as title page

◆ A **detailed abstract** (Figure 7.17) explaining the goals of the research, and serving as an "executive summary"
◆ **Funding justifications**
◆ A **bibliography** containing the publications the researcher wrote (Figure 7.18)
◆ A **table of abbreviations** used in the study (Figure 7.19)
◆ A **reply to preliminary reviewers** of the project
◆ The **proposal** itself, which divides into:
   ◆ **Rationale** for the project
   ◆ **Discussion** of research already accomplished
   ◆ **Detailed explanation** of experimental design
   ◆ **Bibliography** for the whole project
   ◆ **Back matter** (certifications of proper protocol and safety; letters of reference)

Notice that this outline reflects the integration of a proposal or prospectus within the larger framework of a major report. The proposal itself is preceded by writing that validates the authenticity of the writer and subject.

Figure 7.22 depicts some parts of a social services formal report: cover sheet, letter of transmittal, and table of contents. Note how the writer, Pattie Church, has made the beginning of the report "reader-friendly"; though the document discusses an unpleasant problem, she has adopted a positive tone in constructing a cover sheet and setting forth the issue in her letter.

Figure 7.23 displays part of a report by David R. Smith that compares two filtration systems. The document uses tables embedded into the text to create an effective presentation. It employs a covering memorandum, tables of contents and illustrations, an executive summary, and the expected divisions of the longer report. A references page and an appendix displaying credentials were also attached to the presentation, though not reproduced here. This is a document created on the job to achieve a particular purpose.

Figure 7.24 lists tips for integrating figures into long reports, a subject covered more fully in this book's appendix.

Figure 7.14   Preliminary Report by Angela Stewart

DATE:	July 25, 1996
TO:	Technical Writing Peers
FROM:	Angela Stewart *AS*
RE:	Background Report on Ocean and Sea Pollution

INTRODUCTORY SUMMARY

I plan to complete a multisectioned report on ocean and sea pollution. I will thoroughly survey this type of pollution, suggest possible solutions, reveal what I view to be the best solution, and propose a call for action.

Figure 7.14 *(continued)*

In this memo I will explain my research process. I will include the resources that have helped me and those that have not helped. I will also give some of the sources that I still intend to examine.

EARLY RESEARCH PROCESS

I first located several books at the Learning Resource Center that I thought would be helpful. When I actually found the books, however, I could not find any that were written about ocean and sea pollution specifically. There were quite a few books on water pollution; however, all of these were too outdated or too technical. ProQuest was my second, and most helpful, resource. I was immediately able to find ten relevant articles. The information I obtained by using ProQuest was helpful:

- The articles were current—most had been published within the last year.
- A variety of pollution causes were identified and explained.
- Realistic solutions were proposed for ocean cleanup.

The best articles on this topic were as follows:

1. Kowalski, Kathiann M. "Saving Our Beaches." *Current Health 2* May 1996: 18-19. Access NO: 02789612 ProQuest Periodical Abstracts—Research I.
2. Baker, Beth. "Nations Coming to Agreement That Polluted Oceans Need a Cleanup." *Bioscience* Mar. 1996: 183. Access NO: 02732573 ProQuest Periodical Abstracts—Research I.
3. "Salvaging Our Seashores." *Environment* 7 Sep. 1995: 32. Access NO: 02480997 ProQuest Periodical Abstracts—Research I.
4. Brancatelli, Joe. "What Is Polluting Our Beaches?" *Home Mechanix* Mar. 1995: 44-45. Access NO: 02276392 ProQuest Periodical Abstracts—Research I.

At the County Library I found four books on ocean and sea pollution. My only problem was that all of these publications were very outdated—the most recent book was copyrighted in 1981. These books will probably not be helpful to me at all in my research.

CONCLUSION

Next I plan a general search on the Internet. I hope to find the most current data through this system. I also intend to make a final attempt at locating current books on my topic. Finally, I will try to find relevant pamphlets and will check SIRS to see if I can obtain literature pertaining to social implications of ocean and sea pollution.

Figure 7.15   Research Proposal by Angela Stewart

RESEARCH PROSPECTUS

DATE: July 31, 1996
TO: Dr. Brian Holloway
FROM: Angela Stewart *AS*
SUBJECT: Ocean and Sea Pollution

INTRODUCTORY SUMMARY

I plan to complete a multi-sectioned report on ocean and sea pollution. The primary purpose of my report is to inform the reader of the problem; also, I hope to leave the reader with a sense of responsibility and show how each individual can make a difference.

AREAS TO BE STUDIED

My research will involve the following areas:

- **Ocean and Sea Pollution Defined**
  The first section will answer the question: What is ocean and sea pollution? I will clearly explain this type of pollution and how devastating and widespread it has become.
- **Causes of Ocean and Sea Pollution**
  In the second section, I will examine the most prevalent reasons for ocean and sea pollution.
- **Effects of Ocean and Sea Pollution**
  The third section will define the tremendous impact that contamination has on the ocean environment.
- **Positive Ways to Control or Alleviate Ocean and Sea Pollution**
  Finally, I will present the most effective ways of controlling further pollution. I will also share what I view to be the best solution and what the reader can do to help.

METHODS OF RESEARCH

I located several books at the Learning Resource Center, but none were specifically about ocean and sea pollution. There were quite a few books on water pollution; however, all of these were either no longer current or too technical for my intended audience.

The information I found at the County Public Library was very outdated. It may, however, be beneficial in comparing past contamination to that of today. The four books that I found are:

1. Bartlett, Jonathan, ed. *The Reference Shelf: The Ocean Environment.* Vol. 48. No. 6. New York: Wilson, 1977.
2. Loftas, Tony. *The Last Resource: Man's Exploitation of the Oceans.* Chicago: Regnery, 1970.
3. Marx, Wesley. *The Oceans: Our Last Resource.* San Francisco: Sierra, 1981.
4. Moorcraft, Colin. *Must the Seas Die?* Boston: Gambit, 1973.

Figure 7.15   *(continued)*

2

ProQuest has been my most useful resource. Using this system, I immediately found several germane articles. The titles are:

1. Baker, Beth. "Nations Coming to Agreement That Polluted Oceans Need a Cleanup." *Bioscience* Mar. 1996: 183. Access NO: 02732573 ProQuest Periodical Abstracts—Research I.
2. Brancatelli, Joe. "What Is Polluting Our Beaches?" *Home Mechanix* Mar. 1995: 44-45. Access NO: 02276392 ProQuest Periodical Abstracts—Research I.
3. Heinrich, Katherine M. "New Study Traces Padre Island Trash to Shrimpers." *National Parks* Sep. 1995: 21-23. Access NO: 02577190 ProQuest Periodical Abstracts—Research I.
4. Kowalski, Kathiann M. "Saving Our Beaches." *Current Health 2* May 1996: 18-19. Access NO: 02789612 ProQuest Periodical Abstracts—Research I.
5. "Salvaging Our Seashores." *Environment* 7 Sep. 1995: 32. Access NO: 02480997 ProQuest Periodical Abstracts—Research I.

I plan several searches on the Internet and hope to find the most current data through this system. Finally, I will try to find relevant pamphlets, and will check SIRS to see if I can obtain literature pertaining to social implications of ocean and sea pollution.

TIMETABLE

I intend to complete all research by August 1. From August 1 until August 5, I hope to work on rough drafts of my report—August 6 and 7 will be used to polish my work. My completed research paper will be presented to you on August 8.

REQUEST FOR APPROVAL

I request your permission to write my report on the devastating effects of ocean and sea pollution. I would appreciate any input that would benefit my research. I look forward to discussing this topic with you.

Figure 7.16   Formal Report by Angela Stewart

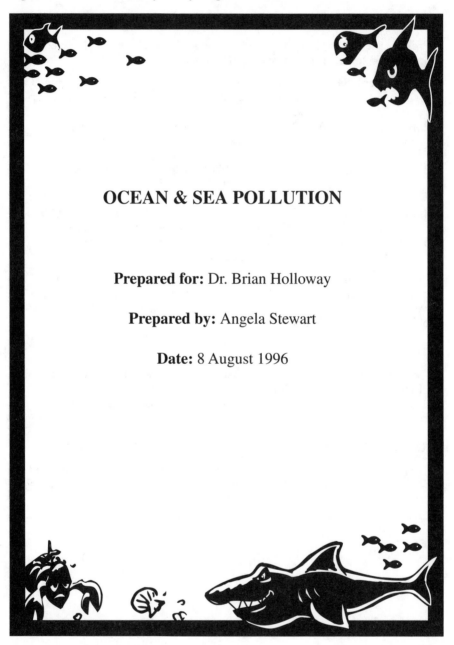

OCEAN & SEA POLLUTION

Prepared for: Dr. Brian Holloway

Prepared by: Angela Stewart

Date: 8 August 1996

Figure 7.16 *(continued)*

MEMORANDUM

DATE: August 8, 1996
TO: Dr. Brian Holloway
FROM: Angela Stewart *AS*
SUBJECT: Ocean and Sea Pollution

Per your request, I have completed a multi-sectioned report on ocean and sea pollution. This report informs the reader of the problem, and demonstrates that each one of us must assume responsibility for the solution.

To create the report, I examined the definition of such pollution and surveyed its usual causes. I then discussed the tremendous impact of contamination on the ocean environment, and presented the most effective ways of controlling further pollution.

I found that my report leaves the reader with the understanding that the problem must be solved by individuals as well as through collective action. Please let me know if you require further information about my project.

Figure 7.16   *(continued)*

Figure 7.16 *(continued)*

EXECUTIVE SUMMARY

This report informs the reader about aspects of ocean and sea pollution and demonstrates that each of us must assume responsibility for correcting this problem.

My project examines the definition and causes of such pollution. It surveys effective ways of controlling further pollution, emphasizing that the problem must be solved by individuals as well as through collective action.

The report consists of four sections:

1. An explanation of ocean and sea pollution and of the problem's general social origins
2. An examination of specific causes of marine degradation
3. A review of the impact of litter and sewage
4. An enumeration of steps to reverse the problem

A source list at the end of the report provides access to further information.

Figure 7.16   (continued)

INTRODUCTION

Miles and miles of shoreline make beaches a precious natural resource. Pristine beaches provide:

1. Pleasure to millions of people every year
2. Support to wildlife ranging from sand dollars to sea grasses
3. A primary economic base for tourism (Kowalski 18)

Unfortunately, many of our beaches are under attack. Pollutants as exotic as giardia—a single-celled protozoan—and as prosaic as the cigarette butt foul our shores, damage fragile ecosystems, and profoundly threaten the simple pastime of spending a day at the beach (Brancatelli 44).

Improvement is unlikely unless immediate action is taken. The coastal population of the U.S. was 80 million in 1990, but will grow to more than 127 million by 2010. It seems inevitable that more people will mean more pollution (Brancatelli 45).

OCEAN AND SEA POLLUTION DEFINED

Much public attention has been given to the degradation of the environment. The most significant changes are those which have occurred in the world's ocean and seas. Nevertheless, it is these negative transitions which have received the least attention (Moorcraft 1).

We still treat the ocean as the endless, infinite pit it was considered to be in medieval times. Expressions like "the bottomless sea" and the "boundless ocean" are commonly used. These sayings reflect the image we have of our oceans and seas—the largest bodies of water on earth (Heyerdahl 147).

Hardly a creek or a river in the world reaches the ocean without carrying a constant flow of nondegradable chemicals—from industrial, urban, or agricultural areas. Directly by sewers or indirectly by way of streams and other waterways, almost every big city in the world makes use of the ocean as mankind's common sink (Heyerdahl 149).

Jacques-Yves Cousteau, French pioneer of underwater exploration, speaks from an impressive array of experience:

> The sea is threatened. We are facing the destruction of the ocean
> by pollution . . . this damage carries on at very high speed—to
> the Indian Ocean, to the Red Sea, to the Mediterranean, to the
> Atlantic. (Moorcraft 2-3)

Figure 7.16 *(continued)*

What we consider too dangerous to be stored under technical control ashore we dump forever out of sight at sea—whether toxic chemicals or nuclear waste. Our only excuse is the still-surviving image of the ocean as a bottomless pit (Heyerdahl 149).

CAUSES OF OCEAN AND SEA POLLUTION

Marine Debris

Approximately 80% of marine pollution comes from human activities on land (Baker 183). The most noticeable type of coastal pollution is "marine debris," or garbage. Debris maims and kills marine life. Discarded fishing line, floats, and lures kill fish, birds, and dolphins. Plastic six-pack holders can cause cuts, strangling, or choking (Kowalski 19). Commercial fishing fleets, military vessels, cruise ships, and pleasure boaters dump tons of waste, plastic fishing gear and nets, and other garbage into the ocean. A survey conducted on the beaches of Amchitka Island in Alaska determined that commercial fishermen left behind almost half a ton of trawl webbing for each mile of beach (Brancatelli 45).

Pathogens

Invisible pollutants, or "pathogens," taint oceans and estuaries. Pathogens reach the sea from raw sewage, sludge, and wastewater, or from storm drains released into coastal waters (Brancatelli 44). These pollutants can present serious health risks. Hepatitis A, E. coli, giardia, and other bacteria can cause:

- gastroenteritis,
- cholera,
- chronic diarrhea, and
- death

to swimmers or consumers of contaminated shellfish or seafood (Kowalski 18).

Nutrient Pollution

"Nutrient pollution," caused by chemicals such as nitrogen and phosphorus, overfertilizes seabeds causing rapid algae and plant growth. The plants then die and decay due to depleted oxygen levels in the water. This leads to mass killing of fish and invertebrates (Brancatelli 45). This type of pollution, called a "brown tide," once wiped out bay-scallop beds in Long Island Sound (Kowalski 19).

EFFECTS OF OCEAN AND SEA POLLUTION

One way that the effect of pollutants can be measured is by beach closures. According to the National Resources Defense Council, 23 states issued almost

3

Figure 7.16 *(continued)*

5,000 beach closures or advisories in 1992 and 1993. Many other unsafe beaches went undetected because states did not have effective monitoring systems (Brancatelli 45). [4]

Litter
- In 1993, the national beach cleanup sponsored by the Center for Marine Conservation collected more than **seven million items of trash**.
- Texas volunteers collected **more than a ton of debris** for every mile cleaned.
- In Connecticut, beach-combers gathered **1,840 cigarette butts** per mile of beach.
- CMC volunteers also gathered:
  - **Over 40,000 rubber balloons**
  - **25,000 plastic six-pack holders**
  - **300,000 glass and plastic beverage bottles**
  - **200,000 metal beverage cans** (Brancatelli 44)

Sewage
- Even though the United States has spent more than $75 billion on wastewater treatment plants in the past 25 years, sewage still flows into oceans. In 1995, untreated waste was flushed into Narragansett Bay, Rhode Island, at the rate of 2.5 billion gallons per year (Kowalski 18).

- During the last ten years, cholera has been found in shellfish beds in Alabama and directly traced to an outbreak in South America. Fortunately, the shellfish beds were closed (Brancatelli 45).

POSITIVE WAYS TO CONTROL OR ALLEVIATE OCEAN AND SEA POLLUTION

What the Government Is Doing

Government is taking action to stop coastal pollution. Continued enforcement of the federal Clean Water Act seeks to stop chemical contaminants (Kowalski 19). In December, 1994, Hanauma Bay Beach in Hawaii became the first "no-smoking" beach (Brancatelli 44).

The United Nations Environmental Program (UNEP) Conference was held last fall in Washington, D.C. More than 100 nations were represented, and a "Global Program of Action" was approved. "It's the first time since the Rio Earth Summit that the nations of the world have been able to come together and focus on the problems of ocean pollution," says Will Martin, Deputy Assistant Secretary of Commerce for International Affairs (Baker 183).

The action plan includes recommendations for reducing marine pollution at local, regional, and national levels. The next step will occur in November of this year. The UN General Assembly will then focus its attention on global ocean pollution (Baker 183).

Figure 7.16   *(continued)*

5

What You Can Do

Everyone can get involved in beach cleanups across the country. These are sponsored by local chapters and affiliates of Keep America Beautiful, the Surfrider Foundation, and other organizations (Kowalski 19). Call **1-800-CMC-BEACH** for information about international coastal cleanup activities.

Action by individuals and groups of citizens can be based on many forms of involvement:

1. As workers involved in processes leading directly or indirectly to marine pollution or overexploitation of a living resource of the seas
2. As inhabitants of houses which pollute the air and water (and hence the seas)
3. As beach-goers, sports fishermen, and sailing enthusiasts
4. As students with the time and facilities to study global and local implications of the decline of the marine ecosystem and to suggest strategies for halting or reversing it
5. As voters in local and national elections
6. As sources of political power of other varieties (Moorcraft 181-182)

CONCLUSION

One unavoidable fact remains: things are going to get worse before they get better. It seems inevitable that a growing population will yield even greater pollution—especially given our dismal record in tending to the aesthetic and ecological damage of beaches and coastal waterways (Brancatelli 45).

This can best be summed up in the words of Wesley Marx—"The predicaments on our watery planet will intensify if we fail to account for our acts in a rational way. We need not exist in a haphazard state of environmental warfare. Like good sailors, we can leave a clean wake" (296).

Figure 7.16   *(continued)*

WORKS CITED   6

Baker, Beth. "Washington Watch: Nations Coming to Agreement That
    Polluted Oceans Need a Cleanup." *Bioscience* Mar. 1996: 183. Access
    NO: 02732573 ProQuest Periodical Abstracts-Research I.

Brancatelli, Joe. "What Is Polluting Our Beaches?" *Home Mechanix* Mar. 1995:
    44-45. Access NO: 02276392 ProQuest Periodical Abstracts-Research I.

Heyerdahl, Thor. "How to Kill an Ocean." *The Reference Shelf: The Ocean
    Environment.* Ed. Jonathan Bartlett. Vol. 48. No. 6. New York: Wilson, 1977.
    147-49.

Kowalski, Kathiann M. "Saving Our Beaches." *Current Health 2* May 1996:
    18-19. Access NO: 02789612 ProQuest Periodical Abstracts-Research I.

Marx, Wesley. *The Ocean: Our Last Resource.* San Francisco: Sierra, 1981. 296.

Moorcraft, Colin. *Must the Seas Die?* Boston: Gambit, 1973. 1-3; 181-82.

Figure 7.17   Abstract: Part of Proposal by Douglas M. Burns

Reports from our laboratory and from several other active groups strongly
suggest that the neuropeptide, kalrectin gene-related peptide (KGP), stimulates osteoblasts. Additional reports demonstrate osteogenic actions and in
vivo skeletal effects of KGP. KGP is abundant in skeletal areas and bone marrow stromal regions, where it is thought to act as a neuroeffector in the mat-
uration and physiologic regulation of osteoblastic cells. We, and several other
well-respected international laboratories, initially demonstrated direct stimulation of osteoblast proliferation and direct stimulation of cAMP production in
osteoblasts at 500-fold lower levels than those necessary to produce effects
in osteoclasts. Bernard and Shih reported that exogenous KGP is strongly
osteogenic in cultures of isolated bone marrow stem cells (data that we have
replicated). Vignery and coworkers have demonstrated that when nominal
quantities are injected into rat, KGP prevents ovariectomy-induced decreases
in tibial bone volume through stimulating osteoblastic action (substantial
increases in the osteoblastic surface, osteoid volume, and mineralization rate).

Our initial studies support a biological role for KGP in skeleton by demonstrating rapid and direct cellular actions on cultured osteosarcoma and
osteoblastic cells that appear independent of adenylate cyclase activation.
Our data suggest that KGP coactivates 3 separate signaling pathways in

Source: Section of Formal Proposal by Dr. Douglas Burns, 1997

Figure 7.17   *(continued)*

osteoblasts, most notably novel activation of membrane ATP-dependent K+ (Katp) channels and resultant attenuation of calcium ion-uptake. Transduction of the KGP signal rapidly results in specific changes in cellular levels of mRNAs encoding both metabolic/signaling and phenotypic osteoblastic proteins important to bone formation. We have now focused on testing whether Katp channel activation or modulation of cytosolic [calcium ion] is responsible for KGP's stimulation of bone sialoprotein (BSP) and procollagen I(alpha) (Col I) gene expression, markers of osteoblastic function. Neither CT nor forskolin replicate this stimulation of BSP and Col I gene expression, but this effect is mimicked by the antihypertensive drug pinacidil, a specific activator of Katp channels.

Based on our initial studies, we postulate that: (1) KGP's activation of Katp channels is a direct action central to all its osteoblastic effects; (2) KGP-induced stimulation of BSP and Col I gene expression is mediated either (a) by activation of membrane Katp channels and attenuation of cellular calcium ion-uptake, or (b) by release of intracellular calcium ion.

To test these hypotheses, we will use rat neonatal osteoblast cultures, and: (a) analyze KGP-induced changes in cellular calcium ion and K+ utilization and membrane potential via fluorescent single-cell imaging and by pharmacological/Rb+1-flux studies; (b) correlate KGP's stimulation of BSP and Col I gene expression with 3 distinct and specific cellular effects produced by KGPs in osteoblasts.

This basic work has direct clinical relevance. Treatment and management of osteoporosis, a debilitating and morbid disease of both aged men and women, is a major healthcare problem, whose estimated 1992 annual cost was in excess of $12 billion. Thirty-three percent of all hip fractures, 40% of forearm fractures, and 13% of vertebral fractures are suffered by males. With an increasingly elderly average population, the incidence of age-related forms of osteoporosis is progressively increasing; it is now estimated that the lifetime hip fracture risk is 15-30% for women and 7-10% for men.

It is difficult to effectively treat osteoporosis when the root cause as well as many aspects of skeletal pathophysiology remain unknown. When the complexities of normal osteoblastic cell regulation are elucidated, it will then be possible to design better therapies. These proposed studies focus on novel regulation of osteoblasts by a skeletal neuropeptide, but they will also broaden our understanding of normal osteoblast biology and cell physiology. Since a key defect underlying metabolic bone diseases such as osteoporosis appears to be a defect in recruitment of mature osteoblastic cells to bone resorption sites, assessment of osteoblastic regulatory peptides may lead to therapeutic peptide development or to novel peptidomimetic therapies.

Figure 7.18 Researcher's Bibliography (Partial)

1. Burns DM, Rodi CM, Agris PF: The natural occurrence of an inhibitor of cell growth in normal and tumorigenic cell lines. *Cancer–Biochemistry and Biophysics* 1:269-277, 1976

2. Burns DM, Touster O: Purification and characterization of rat liver glucosidase II. *Journal of Biological Chemistry* 257:9991-10001, 1982

3. Miller RE, Pope SR, DeWille JD, Burns DM: Hydrocortisone increases and insulin decreases the synthesis of glutamine synthetase in 3T3-L1 adipocytes. *The Journal of Biological Chemistry* 258:5405-5413, 1983

4. Birnbaum RS, Mahoney WC, Burns DM, O'Neil JA, Miller RE, Roos BA: Identification of procalcitonin in rat medullary thyroid carcinoma cells. *The Journal of Biological Chemistry* 259:2870-2874, 1984

5. Miller RE, Burns DM: Regulation of glutamine synthetase in 3T3-L1 adipocytes by insulin, hydrocortisone and cAMP. *Current Topics in Cellular Regulation* 26:65-78, 1985

6. Burns DM, Bhandari B, Short JM, Sanders PG, Wilson RH, Miller RE: Selection of a rat glutamine synthetase cDNA clone. *Biochemical Biophysical Research Communications* 134:146-151, 1986

7. Bhandari B, Burns DM, Hoffman RD, Miller RE: Glutamine synthetase mRNA in cultured 3T3-L1 adipocytes: Complexity, content, and hormonal regulation. *Molecular and Cellular Endocrinology* 47:49-57, 1986

8. Miller RE, Burns DM, Bhandari B: Hormonal regulation of glutamine synthetase in 3T3-L1 adipocytes. *Biology of the Adiposyte: Research Approaches,* G. Hausman, R. Martin, eds, Van Nordstrand Reinhold Co., New York, 1987, pp 198-228

9. Burns DM, Birnbaum RS, Roos BA: A neuroendocrine peptide derived from the amino-terminal half of rat procalcitonin. *Molecular Endocrinology* 3:140, 1989

10. Burns DM, Forstrom JE, Friday KE, Howard GA, Roos BA: Procalcitonin's amino-terminal cleavage peptide (N-procalcitonin) is a bone-cell mitogen. *Proceedings of the National Academy of Science (USA)* 86:9519-9523, 1989

11. Howard GA, Liu C, C. Burns DM, Roos VA: In vivo bone metabolism effects of N-proCT, a novel peptide from the calcitonin gene. *Fundamentals of Bone Growth: Methodology and Applications,* Dixon AD, Sarnat BG, eds. CRC Press, Boca Raton, FL, 345-351, 1991

Source: Section of Formal Proposal by Dr. Douglas Burns, 1997

Figure 7.18 *(continued)*

12. Burns DM, Hill EL, Edwards MW, Forstrom JW, Liu CC, Howard GA, Roos BA: Complementary Anabolic Skeletal Action of CT Gene Products. *Osteoporosis 1990*, C. Christiansen and K. Overguard, eds., Osteopress ApS, Copenhagen, 1301-1307, 1991

13. Edwards MW, Forstrom JW, Burns DM, Roos VA, Howard GA: N-procalcitonin: Kinetics and biodistribution in intact mice, and effect on cortical bone and bone resorption in oophorectomized mice. *Osteoporosis 1990*. C. Christiansen and K. Overguard, eds., Osteopress ApS, Copenhagen, 396-399, 1991

14. Burns DM, Howard GA, Roos BA: An assessment of the anabolic skeletal actions of the amino-terminal peptides from the precursors for calcitonin and calcitonin gene-related peptide. *Annals of the New York Academy of Science* 657:50-62, 1992

15. Kawase T, Howard GA, Roos VA, Burns DM: Diverse actions of calcitonin gene-related peptide on intracellular free Ca-2+ concentration in UMR-106 osteoblast-like cells. *Bone,* 16:379S-384S, 1995

16. Kawase T, Orikasa M, Ogata S, Burns DM: Protein tyrosine phosphorylation induced by epidermal growth factor and insulin-like growth factor-I in the rat clonal RDP-4.1 dental pulp-cell line. *Arch Oral Biol* 40:921-929, 1995

17. Kawase T, Ogata S, Orikasa M, Burns DM: 1,25-Dihydroxylvitamin D3 promotes prostaglandin E–induced differentiation of HL-60 human premyelocytic cells. *Calcified Tissue International* 57:359-366, 1995

18. Kawase T, Howard GA, Roos BA, Burns DM: Calcitonin gene-related peptide inhibits net transmembrane Ca-2+ uptake in osteoblastic cells through cAMP-independent activation of ATP-sensitive membrane K+ channels. *Endocrinology* 137: 984-990, 1996

19. Kawase T, Oguro A, Orikasa M, Burns DM: Characteristics of NaF-induced differentiation of HL-60 cells. *J Bone Min Res,* 11:1676-1687, 1996

20. Kawase T, Howard GA, Roos BA, Burns DM: Acute inhibition of Ca-2+ uptake in osteoblastic UMR-106 cells by parathyroid hormone and prostaglandin E2: Comparison to the effects of calcitonin gene-related peptide. *Endocrinology,* 1996

21. Kawase T, Howard GA, Roos BA, Burns DM: Nitric oxide stimulates osteoblast-mediated in vitro mineralization. *Bone,* 1996

Figure 7.19 Table of Abbreviations

Abbreviations Used in This Narrative

KGP—the neuropeptide "kalrectin gene-related peptide"

CT—the systemic hormone calcitonin

PTH—parathyroid hormone

VIP—vasoactive intestinal peptide

NPY—neuropeptide Y

K_{atp}—ATP-sensitive potassium channel(s)

KCa—$Ca2^+$-dependent potassium channels

Pin—pinacidil, a specific activator of Katp

Gly—glyburide (glybenclamide), a specific activator of Katp

TEA—tetraethylammonium, a selective inhibitor of K_{ca}

PKC—protein kinase C (Ca^{2+} – dependent protein kinase)

PMA—phorbol 12-myristate 13-acetate, an activator of many PKC isoforms

VDC channels—voltage-dependent Ca^{2+} membrane channels

Diltiazem—specific inhibitor of L-type VDC channels

bis-oxonol—the potential sensitive dye, (bis-(1,3-diethylthiobarbituric acid)-trimethine oxonol)

TFA—trifluoroacetic acid

FK—forskolin, an activator of adenyl cyclase

[Ca2+]i—concentration of intracellular calcium ion

BSP—bone sialoprotein

Col I—short for the alpha(1) chain of collagen I

OPN—osteopontin

ON—osteonectin

iNOS—inducible nitric oxide synthase

GS—glutamine synthetase

AR-S—the calcium-specific dye Alizarin Red-S, used in biomineralization assays

TGF-beta—transforming growth factor type beta

IBMX—isomethylbutylxanthin, an inhibitor of cellular phosphodiesterases

Obch—porin-like osteoblastic channel protein

GAPDH—glyceraldehyde 3-phosphate dehydrogenase

A1—nonregulated constitutive mitochondrial protein of unknown function

8-BcA—8-bromo-cAMP, a stable cAMP analog

BAPTA—short name for a complex compound which acts as a Ca2+ sponge

PSS—physiologic sterile saline solution

TCM—tissue culture medium

Source: Section of Formal Proposal by Dr. Douglas Burns, 1997

Figure 7.20 A Graph Helps Readers Understand the Time Line of Experiments Proposed in the Report

Activity	Year 1	Year 2	Year 3	Near Future
A. Single-cell imaging studies				
Study of cell Ca^{2+}	███████████	████		
Study of cell K^+	███████████	████		
Bis-oxanol assay of Em hyperpol.	███████████	████		
B. mRNA study & correlation of regulation with known KGP signaling				
Cellular levels dose-response time-course	████████████	██████		
Test effect of cellular K^+ and Em hyperpolar		██████████	██████	
Test effect of cellular Ca^{2+}		██████████	████	
Retest effects of cellular cAMP		██████████	████	
Extend mRNA findings to study of transcriptional activation			███████	████████

Source: Section of Formal Proposal by Dr. Douglas Burns, 1997

Figure 7.21   A Graph Helps Readers Visualize Material in the Main
Part of the Report

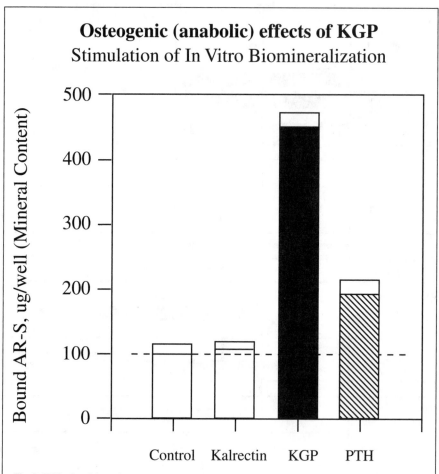

Osteogenic (anabolic) effects of KGP
Stimulation of In Vitro Biomineralization

Bound AR-S, ug/well (Mineral Content)

Control   Kalrectin   KGP   PTH

Fig. 5: KGP stimulates mineralization in osteoblastic cells.

For these quantitative assays, cells cultured in BGJb media supplemented with 1% fbs were treated for
25 days with the indicated biological agent (media changed every 4 days). At the end of day 25, cells
were fixed with 70% ice-cold ethanol (60 min). The ethanol is aspirated, the cells are rehydrated with
water (10 min), and then the cultures are stained with 40 mM Alizarin Red-S (AR-S) for 10 min. Cultures
are washed extensively with PBS to remove nonspecifically bound stain, and then bound dye is solubilized
with cetylpridium chloride (10% in 10 mM sodium phosphate. pH 7.0). Aliquots and dilutions are read in
spectrophotometer and the absorbance at 562 nM read against an AR-S absorbance standard curve for
quantification. The bars indicate the total amount of AR-S specifically bound to the calcified extracellular
matrix in these cultures (mean ±SD; n = 3 dishes per treatment).

Source: Section of Formal Proposal by Dr. Douglas Burns, 1997

Figure 7.22   Sections of Formal Report by Pattie Church

CREATING A DIFFERENCE FOR

HANDICAPPED CHILDREN AND THEIR FAMILIES

LIVING IN A FRUSTRATING WORLD

Prepared for: Concerned Citizens
                 Beckley, West Virginia

Prepared by: Pattie Church
                Advocate for
                Special Needs Children

Date: March 4, 2001

Figure 7.22 *(continued)*

Pattie S. Church
Advocate for Special
Needs Children
P.O. Box 24
Strawn, WV 24555

March 4, 2001

Concerned Beckley Citizens
General Delivery
Beckley, WV 25801

Dear Concerned Beckley Citizens:

My eighteen years of experience working in a local day care facility made me aware of a serious need for child care services designed to accommodate handicapped children and their families in the Beckley area.

Communicating with handicapped children and their families can be extremely distressing. But understanding the overwhelming problems experienced by these families can also inspire people to become advocates for their rights. The purpose of this report is to stimulate public awareness and increase the number of voices heard in favor of creating early intervention methods for these families through proper care facilities.

I encourage you to educate yourselves through the information provided in this report and join the forces of human rights for handicapped children.

Sincerely,

Pattie S. Church

Pattie S. Church
Advocate for Special
Needs Children

Figure 7.22 *(continued)*

Figure 7.23 Filtration Report by David R. Smith

YMCA POOL FILTRATION STUDY

PREPARED FOR: Gary Jones
                                 CEO, BERKLEY YMCA

PREPARED BY: David R. Smith
                               AQUATICS DIRECTOR, BERKLEY YMCA

DATE: MAY 6, 1998

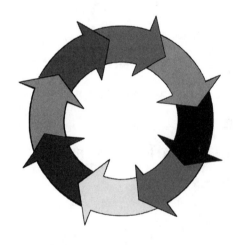

Figure 7.23 *(continued)*

MEMORANDUM

TO: Gary Jones
FROM: David R. Smith
DATE: May 6, 1998
RE: YMCA Pool Filtration Study

As requested, I have conducted a study of two different pool filtration systems (High Rate Sand Filter and Diatomaceous Earth Filter) and determined the best system for the facility. This report will show my research and provide my conclusion as to which filtration system rates highest.

To achieve my results, I had to look at many variables including cost, performance, and cleaning ease. Overall, the Diatomaceous Earth Filter is a little more expensive than the High Rate Sand Filter but for the outstanding service this unit provides it would pay for itself in no time.

Thank you for giving me the opportunity to conduct this study and hopefully better our organization in the process. Feel free to call me at extension 28 to ask any questions that I have failed to answer.

Figure 7.23 *(continued)*

TABLE OF CONTENTS

APPENDICES

LIST OF ILLUSTRATIONS

Figure 7.23   *(continued)*

EXECUTIVE SUMMARY

This study to find out which filtration system would better suit our company
led to discovering many interesting facts about both systems. Although the High
Rate Sand Filter was a little less expensive it needed a lot more attention to
operate and maintain than the Diatomaceous Earth Filter, and the performance
of the High Rate Sand Filter was not up to our standards.

The tests that were conducted to reach the final outcome were determining the
cost of both filters, establishing the performance rate of the machines and
deciding which system would be less of a burden to clean.

These were the results of the tests:

- The average cost of a Diatomaceous Earth filter is $944.60.
- The average cost of a High Rate Sand Filter is $576.20.
- The performance of the Diatomaceous Earth Filter was more efficient
  on an eight-hour turnover rate than the High Rate Sand Filter.
- The Diatomaceous Earth Filter is much easier to clean during and after
  backwashing, and uses less amounts of its cleaning product
  (Diatomaceous Earth).
- The High Rate Sand Filter uses a large quantity of sand and must be
  replaced often.

The study I conducted came to a decisive conclusion and the Diatomaceous
Earth filter came out the clear winner as the best system for our organization.

I

Figure 7.23   *(continued)*

INTRODUCTION

The document to follow is a study conducted for the Berkley YMCA Aquatic Facility to determine the best pool filtration system to purchase for the business. So you can better follow this report, I have included a brief description of the study, a summary of how I went about gathering information and actually doing the research, and an overview of the report's format.

Project Description
Six months ago the YMCA aquatics staff and maintenance personnel started having a lot of trouble with the swimming pool filtration system and as a result spent countless days trying to repair and salvage the old system. One week after this problem came about I was asked to conduct a study of the Diatomaceous Earth Filter and the High Rate Sand Filter and then determine which system would be efficient and job worthy at the YMCA. I chose three main topics to look at to determine the best filter: (1) cost of both filters, (2) performance of the filters during an eight hour turnover rate, and (3) the ease of cleaning or backwashing the filters.

Scope of the Report
A lot of my resources derived from personal interviews, as well as personal knowledge from being a Certified Pool and Spa Operator. I also looked on the Internet and found a few good articles on the two filtration systems. After doing my research, I was able to come to my conclusions and recommendations based on my findings.

Report Format
This report is composed of four main sections:

- Findings on the cost of both filters
- Findings on the performance of the systems
- Findings on the ease of cleaning the filters
- Conclusions and recommendations

An appendix at the end of the report contains a copy of my Certified Pool and Spa Operator's certificate.

2

Figure 7.23 *(continued)*

ACTUAL STUDY

High Rate Sand Filter Overview
High Rate Sand Filters were invented in the 1950s and are still in use today.
They use fine sand particles to filter dirt and other particles out of the water.
Water is circulated from the pool into the filter, at which time the water is
dispersed over the sand, trapping dirt and other particles as the water is filtered
through the sand and back into the pool.

Cost
The cost of High Rate Sand Filters is relatively low as can be shown in the
table below:

TABLE 1

MODEL NUMBER	TYPE OF FILTER	LIST PRICE
S200-SIDE MOUNT	HIGH RATE SAND	$691.00
S210S-SIDE MOUNT	HIGH RATE SAND	$620.00
S244S-SIDE MOUNT	HIGH RATE SAND	$711.54
FLT-S180T18"	HIGH RATE SAND	$390.50
FLT-S210T20"	HIGH RATE SAND	$469.00

The average cost of the High Rate Sand Filter is $576.20 (Earthworks 2).

Performance
The High Rate Sand Filter has an average performance rate. Although it can
filter small particles out of the water, if the sand is not changed regularly the
water will cause channeling and this will induce the dirt and other particles to
return to the pool causing turbidity or cloudiness of the pool. The High Rate Sand
Filter also has a high flow rate, and uses more electricity to operate. The flow
rate of a High Rate Sand filter is 12 to 20 gallons per minute (Ingrassia 15).

Cleaning
Backwashing is the best way to clean the High Rate Sand Filter. This process
involves reversing the flow of water to the filter, forcing the dirt and debris to be
pushed out of the sand and into a drainage line, which discards the waste. The
rule of thumb for cleaning a High Rate Sand Filter is about once a week
(Ingrassia 17).

3

Figure 7.23 *(continued)*

Diatomaceous Earth Filter Overview
A Diatomaceous Earth Filter uses Diatomaceous Earth or Diatomite as the filtration medium. D.E. is the fossil remains of tiny aquatic plants called diatoms that settled at the bottom of the ocean floor. D.E. particles resemble snowflakes, are porous, and are measured in microns (Diatomaceous Earth (DE) Filters 1). The Diatomaceous Earth is placed on fabric elements then the water is forced through the elements and D.E. causing the dirt, grease, blood cells, or anything else that passes through it to become trapped in the Diatomaceous Earth-covered elements.

Cost
A Diatomaceous Earth Filter is more expensive than a High Rate Sand Filter. The table below shows the costs of five Diatomaceous Earth Filters.

TABLE 2

MODEL NUMBER	TYPE OF FILTER	LIST PRICE
EC65A-PERFLEX	D.E.	$719.00
EC75A-PERFLEX	D.E.	$814.00
DE6000A	D.E.	$1236.85
DE4800A	D.E.	$1019.00
DE3600A	D.E.	$935.00

The average cost of a Diatomaceous Earth Filter is $944.60 (Earthworks 2).

Performance
The Diatomaceous Earth Filter has an outstanding performance rate. The D.E. Filter can remove the smallest particles (even red blood cells) better than any filter, including the High Rate Sand Filter. The Diatomaceous Earth Filter also has a considerably lower flow rate than the High Rate Sand Filter, making the D.E. Filter more economical. The flow rate for the Diatomaceous Earth Filter is two gallons per minute (Ingrassia 16).

Cleaning
Cleaning Diatomaceous Earth Filters can be done very easily by hand rinsing with a regular water hose. You can also backwash the filter if the Diatomite does not come off of the fabric entirely. When you are finished cleaning the filter you have to coat the elements with fresh Diatomite. Simply adding a new bag of Diatomaceous Earth to the filter does this. As the water is circulated throughout the filtration system the Diatomite re-attaches to the fabric elements. This is called caking (Ingrassia 18).

4

Figure 7.23 *(continued)*

CONCLUSIONS AND RECOMMENDATIONS

These are the final conclusions and recommendations concerning the study I have done for the YMCA to determine the best pool filtration system for the organization.

Conclusions

Generally the High Rate Sand Filter and the Diatomaceous Earth Filter are the most commonly used filters throughout the United States, and both are relatively good. However, because of the size of the YMCA swimming pool and the number of swimmers who occupy the pool daily, I have come to the conclusion that the Diatomaceous Earth Filter would be the best filter for our organization to purchase. The Diatomaceous Earth Filter can filter parti-cles to three (3) microns which is about the size of a red blood cell, and the filter doesn't need as high a flow rate as the High Rate Sand Filter which would save the YMCA money on electricity bills. Finally the Diatomaceous Earth Filter does not use as much Diatomite as the High Rate Sand Filter uses sand, which cuts down on the overall cost of the filtration system.

Recommendations

After examining these conclusions I have made the following recommenda-tions for the YMCA pool facility:

1. Discard the current filtration system before it causes serious problems for our facility.
2. Purchase a Diatomaceous Earth Filter to be used in place of our current filtration system.
3. Advise pool maintenance personnel to conduct regular tests and mainte-nance to ensure the long life of the system.

Following these recommendations, the YMCA pool will be a much cleaner and more attractive place for the people of Berkley who wish to enjoy our facility.

5

Figure 7.24   Integrating Art into Reports

Often, reports benefit from artwork—also called "graphics" or "visuals." You may decide to insert a visual as

- A closing to a section, where it will provide emphasis and clarity
- An explanation within a section, used to help readers grasp a concept
- Back matter in an approach displaying data

Attaching graphics to the appropriate spot in the report may be done with better grades of clear adhesive tape. Photocopying the page with the attached visual then produces a clean final copy.

As an alternative, position the picture on a blank page, copy that, then arrange the text in your word processor to leave a gap at the location where the graphic will be. Print that text over the already-copied page for a crisp result.

If you have an optical scanner, of course, just capture the image and insert it directly into the report!

Assignments

Examine Figures 7.14–7.23 to answer Questions 1–3.
1. What strategies of writing help make these reports coherent—that is, help glue the parts together for effective transactional writing?
2. How do subordinate features of these documents provide support for main concepts?
3. What audience-relation techniques do you notice?
4. Outline the organizational plan of Angela Stewart's final project (Figure 7.16), identifying material in the paper that first appeared in her preliminary report and proposal.
5. Study the ISO 9000 report in Chapter 6 (Figure 6.10). Compare it to the reports shown in this chapter. What could be done to the ISO report to make it "formal"?
6. Study the research proposal shown in Figure 7.25. Here, Aimee Salmons presents a prospectus for a multi-sectioned paper on a problem in public education. Because she has planned her task carefully, the proposal's introductory summary will develop into the "message" material of her long report, the questions to be answered will evolve into the "support" sections, and the description of research will be expanded into a reference list or bibliography. Explain how this will be accomplished—
   a. What must be added to the introductory material, and how must such material be divided into different parts?
   b. How might the parts of her hypothetical introduction reflect her mission without repeating themselves?

c. What subordinate material might be organized under which main headings in the body of the report?

d. What features must the final report's reference list have that are not present at this stage of the project?

Figure 7.25 Proposal for a Multi-sectioned Report by Aimee Salmons

---

TO: Dr. Brian Holloway
FROM: Aimee Salmons
DATE: November 12, 1997
RE: Research Proposal

Introductory Summary
The purpose of my formal report will be to examine the effectiveness of current sexual education programs in schools and their communities. I also intend to focus on the significance of peer pressure, inquisitiveness, and in-experience in shaping a teenager's attitude towards sex. My thesis will read as follows:

"In order to fight such powerful enemies as peer pressure and curiosity, we have to show our children how to do what's right for them and how to know *when* it's right."

Questions to Answer
In this report, I would like to address the following questions:

• Is teaching abstinence as the only solution for teens effective?
• Are sex education programs sending mixed messages to students?
• Are the students more receptive to dual messages?
• Should the government be more involved in program development?
• Are existing programs being tested properly to see if they work?

Methods of Research
Since sexual education in schools is primarily a social issue, I have been relying on SIRS for most of my research. The main body of my preliminary research comes from these three articles:

Buckley, Stephen; Wilgoren, Debbi. "Young and Experienced." *Washington Post* April 24 1994: A1+. SIRS 1994 Youth, Volume Number 4, Article 67.

Roan, Shari. "Are We Teaching Too Little, Too Late?" *Los Angeles Times* July 12, 1995: E1+. SIRS 1995 Youth, Volume Number 4, Article 93.

Daley, Daniel; Wacker, Betsy. "Sexuality and the 104th Congress: The First Hundred Days." *SIECUS Report* June/July 1995: 13-15. SIRS 1995 Sexual-ity, Volume Number 4, Article 52.

Figure 7.25 *(continued)*

2

I also conducted searches on the Internet as well as ProQuest, but did not discover any articles relating to the exact material to be discussed in my report. When I attempted to search for books in the Learning Resource Center, I could not find any current data on sex education programs or their effectiveness.

Timetable
I plan to complete my research by December 1 and finish writing my report by December 5. I then expect to have a cover sheet, letter of transmittal, table of contents, and any "back matter" completed two days before the final examination date for this class.

Request for Approval
I have now shown you what questions need to be asked about sexual education in schools, and I am also forming some hypothetical solutions to this problem with the use of my information. I hope you approve of my topic choice and allow this study to continue so I can present my information to you in the final report. Please inform me if you have any problem with this topic.

8

Selling Yourself

Introduction

Today's world is uncertain, and the employment marketplace mirrors that uncertainty. You may change careers five or more times during your working life—and even within careers, jobs can quickly become unstable. Just as you should bank some money from every paycheck to anticipate future challenges, you must also save information that will help you find a new job. And for many, the job-seeking experience is the ultimate in transactional writing—calling for the persuasive, informative, and research strategies we have surveyed. In addition, documents produced during the job search must adhere scrupulously to details of layout and format.

Keeping a Work-History File

The information you must retain to anticipate the sudden necessity of finding a job should be kept in a work-history file: a continuously updated archive of your employment. Use manila file folders and label them as follows:

◆ Employment History (past jobs)
◆ Current Job
◆ Volunteer Work
◆ Other Experience and Skills
◆ Honors and Awards
◆ References
◆ Current Résumé

Then supply the information needed for these folders. Keep an accurate, descriptive list of all past employment—perhaps categorizing it by skills (analytically) rather than chronologically (by narrative), especially if you have shifted fields frequently. Sorting your work experiences both ways will assist you in creating chronological and functional résumés later. Write down what your responsibilities were at each position, and note what you learned from each job. Do the same for your current position. List and describe volunteer work (PTA, Scouts, church, etc.) and the skills you used in these volunteer capacities. List other skills you may have that have not been exercised in volunteer work or employment. Record any honors or awards you have received, and maintain a current list of references with corresponding telephone numbers and addresses both postal and electronic (you'll want references of ability and character). Finally, you'll need to keep copies of your current résumé handy; we'll examine the art of writing résumés shortly. Remember that even though you have this standard résumé, you may be required to complete a much larger form as a condition of application (for civil-service jobs, for example) or as a condition of employment. Your work-history file will help you devise a résumé, fill out lengthy forms, and respond to interviewers' questions.

Don't delay constructing this work archive—it's like having a compass when you're in the woods. As an old camper once noted, "It's far better to have one and not need one than to need one and not have one!" Avoid the frustration of trying to unearth old records and remember old addresses after the fact. Having all your information organized and ready will expedite your search for new employment.

Exercise

Create your own work-history file.

The Modern Job Search

Today, many companies can choose from a surplus of candidates for employment, and therefore evince interest in only those job seekers who meet specific company needs. Because of this, you can't know *enough* about the firm to which you apply. Therefore, you'll not only need to do the traditional research in the library—studying the organization, goals, and financial features of companies, using reference books and periodicals—but you also will need to adopt contemporary strategies to investigate likely prospects. It is not unusual today for mid-sized and larger firms to post their corporate structures, the names and phone numbers of division heads, their job openings, and even their stock prospectuses on the Internet. And because the total number of Web sites is surging toward the billions, the Internet will continue to be a primary resource for the job hunter. Figure 8.1 presents, however, a few concerns about such directories on the Internet. Certainly, as companies grow and change, these directories will evolve—necessitating frequent returns to their Web sites!

Here are some tips for researching job prospects:

1. *Use initiative.* Realize that often a direct call to the personnel department will generate the information you want.
2. *Use networking.* Do acquaintances, friends, or relatives work for this firm? Can you make contacts at trade and professional conferences?

Figure 8.1 Examining a Corporate Directory on the Internet

The Internet is ephemeral at times; much can change between the time the site is developed and when you refer to that site. Remember that—

- A division newly added to the firm may have been renamed or sold.

- The heads of departments may be different.

- In a rush to present itself visually to a generic audience, the company may not have developed a home page revealing its complete structure or the depth of its organization.

- You should not expect to find a financial report at that site. Supplement all your findings on the Net with material gleaned from "hard copy"— financial reports available from the firm or from brokerages, and the evaluative analysis of reference books in the library.

3. *Be mobile.* Visit those professional and trade conferences, especially if they are within a day's drive and therefore convenient. Make sure your new contacts have your name, phone number, address, and résumé; if appropriate, hand out your business card. Don't wait until after you've graduated to familiarize yourself with your field.
4. *Be knowledgeable.* Learn everything you can about your potential employer. At the trade show, did the employees seem harassed? Were they dressed in worn suits that seemed to have been slept in for three days? Were the organization's plans vague? These are some negative clues. You will need much more than a "first impression" discovered under duress, however; you will need material that only the library and the Internet can supply.
5. *Be Net-conscious.* Use Internet bulletin board services, and post your résumé on the Internet. Use Internet research resources too.
6. *Be accessible,* both by e-mail and phone message service, to take advantage of "impulse" calls by recruiters seeking more information about you.

Traditionally, the reference collections of libraries have been the repositories of books helpful in the job search. Of course, you'll need to examine current collections of model résumés and cover letters—standard fare in most libraries—but you also want to study books such as these, usually found in the business section of the reference collection:

1. *Standard & Poor's Register of Corporations, Directors, and Executives* (discloses structure and background)
2. *Moody's Handbook of Common Stocks* (contains analyses and predictions)
3. *Hoover's Masterlist of Major U.S. Companies* (has sketches of these firms)

4. *Business Information Sources* (is a bibliography)
5. *The National Job Bank* (lists jobs and their locations)
6. *The American Almanac of Jobs and Salaries*
7. *Professional Careers Sourcebook*
8. *Job Hunter's Sourcebook*
9. *Directories of Manufacturers* (these list the firms in different states)
10. *The Job Seeker's Guide to Private and Public Companies*
11. *Occupational Outlook Handbook*

I also recommend reading the current annual edition of *What Color Is Your Parachute?* by Richard Bolles for tips about the job-search process. Though older editions are likely to be found in libraries, the most up-to-date volume will be in bookstores.

Additionally, a quick trip to the front page of many popular Internet providers will afford access to company information directories online, such as Hoover's, as well as pages for career research including, at the date of this writing, Car.Builder, Career.com, CareerPath, HeadHunter, HotJobs, JobOptions, Jobs.com, and Monster. (Because of the volatile nature of the Internet, these sites or their titles may change).

As you can guess from this discussion, most of these documents have self-explanatory titles and are organized for quick reference. Using these resources, and others like them, you can discover the size, income, ownership, hierarchy, structure, history, and financial status of many prospective companies. You can also figure out what types of jobs are available in different geographical areas. Add to that the specific information about companies and open positions that you can acquire by reading the journals and magazines published in your trade or profession—most of which should be available in a good library—and you have a good field of information on which to base your search for likely employment.

If you are a college student or alumnus, you should also visit the college placement office frequently—and be sure to ask if that office has a referral feature distributing résumés to alumni chapters around the country. Also, ascertain whether your placement office subscribes to one of the electronic job-searching databases. Using these databases, companies that search by keywords—such as "marketing + management"—are connected to résumés in electronic format. Some databases provide job-seekers with categorized employment listings. Certain college career centers offer free access to one of these services, or charge a nominal fee. Employment agencies usually make similar services available, and this information can be privately purchased as well. Jobtrak and Skillsearch are examples of the many database tools available, and these as well as others may be researched on the Internet. Your work-history file will be invaluable in completing the forms some databases require. Remember that you should use free services first, since private access to such entities can cost money. These companies have changed the way résumés should look—a subject discussed in the next section of this chapter.

Exercise

Research a prospective employer as if you were seeking a position with that organization.

1. *Begin* by looking through the job postings in trade or professional publications, the Sunday editions of metropolitan newspapers (such as the *New York Times*), Internet advertisements (find them by using subject keywords), and other employment listings.
2. *Select* a likely prospect and photocopy that posting for reference.
3. *Keep* that photocopy in a file folder marked with the job designation; take that folder to the library and add to its contents notes taken on the company based upon your print and electronic research.
4. *Assess* that job's strengths and weaknesses in light of what you have just learned. Would you fit into the organization? What are the typical benefits and salaries? What is the potential for career advancement—is this company expanding or contracting?

Résumés

Gone are the days when résumés could be composed of chatty paragraphs in a list. Gone are the days of colored paper; scented paper; and pictures of the applicant, applicant's family, and family dog. Gone are the days of odd fonts, parallel columns, and bulky, multipage résumés (except in special cases). There are still books in circulation recommending what is bygone—if it ever really was acceptable—but avoid the items mentioned. Today, your résumé may be scanned into a database, and electronic scanners have trouble with columns, peculiar type, and colored paper. Lengthy descriptions have yielded to current keywords—since these are how résumés may be filed and searched, and since an initial reading of a résumé might be limited to five seconds.

In fact, you may post your résumé online from the start. Doing so is beneficial if your content in its paper version exceeds one page, but to make the instant impact you desire, you will need a very strong job objective and listing of skills, as explained below. In addition, forego distracting pictorial material or animation that may annoy the recipient and might make the document harder to send or input. Of course, illustrations of the projects your résumé mentions may be stored as files and linked to the key words of your résumé text, so that the serious recruiter may find and study them; plus, your résumé can then be sent as an e-mail attachment instantly. These aspects are decided advantages to posting your résumé electronically in addition to distributing conventional paper copies.

A résumé is really a "foot in the door"—something to arouse a potential employer's interest. It cannot take the place of a work-history file. Rather, the résumé is a brief, concentrated view of your skills and background. It is, in this sense, a super-summary. Because of differences in emphasis, you will need to construct alternative résumés for different jobs. Résumés are usually organized according to one of three concepts.

Chronological Résumés

Chronological résumés use *narrative* as the basis of organization, emphasizing cumulative accomplishments. They usually display one's work experience in reverse chronology—most recent employment first. Such résumés work well for those who have been developing within one career and for whom the next job is a logical extension of such evolution. They may not be suited for those who have changed jobs or careers frequently, or for those who have substantial gaps in time in their work histories. They may be unsuitable for a college student seeking a job as an electrical engineer who has put herself through school working at fast-food restaurants and part-time sales. Though valuable skills are learned and honed in such jobs, they do not provide the consistency in career orientation characteristic of the best chronological depiction. However, there are exceptions.

Figure 8.2 presents a chronological résumé in which the writer has combined several different work experiences with a multidisciplinary college education; Kathy Latham's "Objective" heading shows how her diverse background makes her a good candidate for a particular job. This was a hard résumé to construct chronologically because of the career changes and gaps that it reflects, but the result is smooth—a good match for a strong cover letter.

Functional Résumés

Because they are organized *analytically*, functional résumés overcome the problem faced by individuals with job shifts and employment gaps. Rather than emphasizing continuity over time, the functional résumé categorizes the applicant's skills. Attributes such as working with people and handling money responsibly might be labeled "Customer Relations" and "Finances," for example. Consider the functional résumé if your jobs have been diverse or your career path has been interrupted. Figure 8.3 shows Kathy Latham's general-purpose résumé, which targets many positions. Though the résumé features skills, it still provides a brief list of employment. Figure 8.4 depicts another functional résumé emphasizing business strengths.

Mixed Résumés

The mixed résumé works well if you wish to display both chronology of employment and key abilities. Remembering that many electronic résumé

Figure 8.2   A Chronological Résumé

Kathy Latham
456 Pell Court
Surrey, Illinois 63098
216-555-5555
klatham555@aol.com

OBJECTIVE   Entry-level management position with educational services corporation requiring skills in marketing and administration

WORK   1995-97 Vice President, Surrey Park District Board
—Budgeting, Contracts, Appropriations
—Educational Programs

1995-97 Teacher's Aide, Surrey District 115, Illinois
—Gifted, General, and Remedial
—Surrey Schools Commendation

1989-95 Volunteer, Early Childhood, Surrey Schools
—Specialized in Computer-Enhanced Learning
—Vice-Chair of PTA

1982-89 District and Store Manager, Oak Farms Foods
—District Manager, 24 Stores, Three States 1983-88
—Store Manager, Surrey Mall, 1982-83
—Highest Profit in Franchise Group Award

EDUCATION 1997-2001   Completing Business Degree at Powell College
—B.S. with Concentration in Management
—Courses Include Management, Accounting, Computer Science and Systems
—Associate of Arts Degree in Teacher Prep

1978-81   General/Technical Courses—University of Missouri

REFERENCES Available upon request

searches proceed by job description and important skills, you might decide to craft a document that exhibits both to advantage. Figure 8.5 exhibits a mixed résumé, splitting the emphasis between skills and work experience. The mixed format works well online, where skills must be immediately apparent and details of work important, and where linking the keywords to other files can enable the interested reader to access further information about you.

Figure 8.3   Functional Résumé

Kathy Latham
456 Pell Court
Surrey, Illinois 63098
216-555-5555
klatham555@aol.com

OBJECTIVE   Entry-level management position with educational services corporation requiring skills in marketing and administration

SKILLS

Management:	Administration
	Financial Evaluation
	Scheduling, Training, Sales
	Inventory, Purchasing
	Hiring, Interviewing
Computer:	Macintosh and PC
	Spreadsheet, Graphics
	Word Processing
	Educational Programs
Office:	Writing/Verbal
	Organizational, Clerical

EDUCATION

1997-2001	Completing Business Degree at Powell College—B.S. with Concentration in Management
1978-81	General and Technical Courses University of Missouri–Columbia

WORK

1995-97	Vice President, Surrey Park District Board
1995-97	Teacher's Aide, Surrey District 115, Illinois
1989-95	Volunteer, Early Childhood, Surrey Schools
1982-89	District and Store Manager, Oak Farms Foods

HONORS   Top Stores Certificates from Oak Farms Foods
Surrey District 115 Commendations

REFERENCES   Available upon request

Figure 8.4 A Functional Résumé

RON BLUFIELD
227 Streamview Lane
Leapton, Illinois 25813
(217) 555-0555
ron.bluefield@here.net

OBJECTIVE
District Manager position enhanced by marketing, administrative, and writing abilities

SKILLS

Marketing
—Began retail business from scratch in 1983
—Business grew to number one in Midwest by 1988
—Strategies commended in trade publications

Administration
—Oversaw development of business from proprietorship to corporation
—Developed sales and distribution organization
—Transferred skills to running college department

Writing
—Created flyers, catalog, promo videos
—Wrote sales manual
—Wrote personnel manual for college department

WORK EXPERIENCE

—Management Consulting 1993-present
—Red Rock College and Parton College—Teaching 1992-present
—MegaMusic (Owner) 1983-1992

EDUCATION

Business and Marketing Seminars, 1983-86, U. of Illinois
Ph.D., Renaissance History, 1985, U. of Illinois

AWARDS/SERVICE

—"Retailer of the Year" awarded twice by Wholesaler's Assoc.
—Published newspaper column, business articles
—Commended in trade journals—*MTR, MMC*

REFERENCES

Available upon request

Figure 8.5   A Mixed Résumé

Kathy Latham
456 Pell Court
Surrey, Illinois 63098
216-555-5555
klatham555@aol.com

OBJECTIVE   Entry-level management position with educational services
corporation requiring skills in marketing and administration

WORK   1995-97   Vice President, Surrey Park District Board
—Budgeting, Contracts, Appropriations
—Educational Programs

1995-97   Teacher's Aide, Surrey District 115, Illinois
—Gifted, General, and Remedial
—Surrey Schools Commendation

1989-95   Volunteer, Early Childhood, Surrey Schools
—Specialized in Computer-Enhanced Learning

1982-89   District and Store Manager, Oak Farms Foods
—District Manager, 24 Stores, Three States 1983-88
—Store Manager, Surrey Mall, 1982-83
—Highest Profit in Franchise Group Award

SKILLS   Management: Administration
Scheduling, Training, Sales
Inventory, Purchasing
Hiring, Interviewing

Computer: Macintosh and PC
Spreadsheet, Graphics, Word Processing
Educational Programs

Office: Writing/Verbal, Organizational, Clerical

EDUCATION   1997-2001   Completing Business Degree
at Powell College—B.S. with
Concentration in Management

1978-1981   General and Technical Courses
University of Missouri

REFERENCES   Available upon request

As with any other type of writing, there are appropriate stylistic conventions that must be followed in creating résumés.

DO

◆ Use phrases rather than long sentences—remember that the résumé may have to be read in five to ten seconds the first time around, and more carefully the next
◆ Keep phrases short—avoid having to drop down to the next line to complete a thought, since that complicates a quick reading
◆ Use parallelism throughout; use active verbs as in this list
◆ Delete anything not essential

DO NOT

◆ Wander or divert the reader's attention
◆ Forgo proofreading—a misspelled word or grammatical atrocity will negate any interest in your résumé
◆ Include information about gender, marital status, children
◆ Discuss height, hobbies, ethnicity
◆ Expand your résumé beyond one page unless such is the accepted practice in your field

Parts of Résumés

Before writing your own résumé, you should study the preceding examples and consider the essentials that belong in the various sections. Although many word-processing programs contain "résumé-maker" software, such software can restrict your ability to present yourself; you may attempt to create a distinctive appearance, only to be told by the computer that you cannot make changes to a locked document. Anyone with a word processor, a computer, or even an electronic typewriter can make excellent résumés without recourse to a prepackaged model, however.

Remember that a good résumé is a work of art—minimalist art, to be sure, but art nevertheless. Use no smaller than 10-point type; 12- or 14-point is best. Use bold for key headings. Follow an outline format as in the examples, including these basics:

1. Your name, address, and telephone number constitute the *heading*. This information should be centered at the top of the page or presented either flush right or flush left. A type font larger than that used in the body of the résumé, or of a slightly different style, will draw attention to it as if it were the title of a report. You might want to experiment with placing a single or double line below your heading for visual definition.
2. The *job objective* acts as the résumé's controlling thesis or statement of purpose; résumés without such a statement may appear unfocused and

generalized. You must tailor this objective to the specific posting for which you apply. Job objectives are usually stated in the following ways:

- "To obtain a [name of job] providing an opportunity to [do whatever the job defines as its requirements]."
- "To use my [specific type of] skills to benefit your company in [name of position]."
- "To obtain a [name of job] using my skills in [field] to [explain how you will enhance company]."
- "Seek [name of job]."
- "Seek [name of job] to [explain benefit for company] using my [describe skills]."

You will need to tinker with the phrasing of this important element—try to say the most with the least verbiage, but be sure that you show a benefit to the potential employer. Note Kathy Latham's brief but accurate example in Figure 8.2.

3. State your *educational experience* appropriately (don't include junior high and elementary school if you're applying for a post-college job). Do include appropriate training received in the military, as well as courses taken for licensure and certification. The most recent experience usually comes first; follow reverse chronology. The writer of either a chronological or functional résumé whose education occurred more than ten years ago may elect to place the "education" section after the description of work experience.

4. *Work experience* follows in a chronological résumé. If the résumé is a functional one, it follows a list of *skills*. In either case, keep descriptive phrases short and parallel; use dashes, bullets, bold type, or extra spacing after lines to emphasize items.

5. A category such as *awards and service, other experience,* or *related experience* sometimes appears as the second-to-last element in a résumé—if needed. Use this section only if it is appropriate—for example, if you read and speak Spanish fluently and are applying for a position potentially involving that ability, you will want your readers to know of your talent. Similarly, if the awards received are relevant to the job sought, mention them—but then avoid including them in another category ("skills," for example).

6. It is customary to close a résumé with a statement that references are available upon request. If you have the space, these may be added at the bottom of your résumé, though this is no longer the usual practice. Instead, make a separate sheet labeled "References" to give to interested parties, and consider supplying actual letters of reference with that reference sheet. If you're running out of room on your résumé, just delete the "References" line entirely so that you can emphasize important content.

Exercise

Create a chronological, a functional, and a mixed résumé targeting the job you have been researching. To acquire further background, look at résumé examples posted on the Internet as well as in reference manuals in the library. Place a copy of each résumé in the file containing the description of and the notes about your targeted job. Assess your résumés—which one best reflects your skills, work experience, and benefit to the company? Might each type of résumé work better in different situations? Why?

Cover Letters

Cover letters and résumés are normally sent together when one applies for a position. Detail and focus are crucial in constructing such letters, which must develop explanations of key skills and accomplishments as well as present an organized overview of your qualifications. Such letters should be well-crafted persuasive documents adhering to the message-support-closure pattern. Remember these points about cover letters:

1. *Focus* your cover letter by incorporating a thesis or a statement of purpose in the very first paragraph. This beginning paragraph should explain why you are writing and where you saw the job posting. Perhaps you haven't seen an ad at all, but are sending a letter of inquiry; if so, state this. Here are some examples of "message" paragraphs:

   "I write to apply for the position of Sales Manager posted in the May 12 *Chicago Tribune*." [Statement of Purpose]

   "I write to apply for the job of Sales Manager posted in the May 12 *Chicago Tribune*. My marketing and merchandising expertise will benefit your company." [Purpose Combined with Thesis]

   "My advisor in ceramic engineering, Dr. Thornton Smith, suggests that I contact you about a new opening in your materials research lab. I wish to apply for the position." [Statement of Purpose]

   "I write in reference to the research assistant opening that my advisor, Dr. Thornton Smith, tells me will soon be advertised. My experience running a small lab and my academic work qualify me for this position." [Purpose and Thesis]

   "I am responding to the advertisement for a social worker–children's advocate, which was posted in the December 6 *Times*." [Statement of Purpose]

   "I am applying for the child welfare specialist opening posted in last week's *Times*. My field experience, degree, and commitment to children will benefit your agency." [Purpose and Thesis]

   Notice that those first paragraphs that are statements of purpose *only* may lack the focus provided when a thesis is also used. Note as well that a thesis that forecasts—or itemizes—obligates you to discuss those items. This is good: your letter will have structure and your reader will know exactly where the letter is going. If you use a thesis, it should take the general form of "My work experience and my education qualify me for this job." Such a thesis usually signals to the reader that there will be two paragraphs in the "support" section of the letter—one about work experience and one about education. Either of these two subjects can be treated first, depending upon their order in the forecasting thesis.

2. *Audience relationship* is crucial. Do use the buzzwords appropriate to the field but don't try to impress the reader with your command of sesquipedalian synonyms. Teachers in adult-education programs are not usually referred to as "pedagogues externalizing andragogical models," for example. Also, strike a forthright tone. You *do* want the job, so convey that desire through straightforward language. Never adopt a posture of superiority, as in the phrase "as you can see if you will just look at my résumé …"

3. *Organize* the body to follow your thesis. Within the organization, develop important points but don't try to regurgitate the résumé. Focus on the most important reasons why the company should hire *you*. Generally, as we've seen, the body of a cover letter divides into two paragraphs, one discussing work experience and the other discussing education. Exceptions occur when the two subjects are intertwined—co-ops and internships, for example—or if both are short.

4. *Close* the letter by inviting further communication. Explain that you can be contacted by mail at a certain address and by telephone at a stated number. Include specific times that are best to call. Supply your e-mail address. Express the desire to hear from your recipient soon.

5. *Size* is important. Study sample letters in your field, found in anthologies of résumés and cover letters. There is no fixed page requirement, but there had better be a very good reason for a cover letter to exceed one page. Your cover letter should target and tantalize, but it cannot and should not replace the more thoroughgoing exchange of information that occurs during an interview.

Examples of cover letters follow in Figures 8.6 and 8.7. Figure 8.6 contains a letter that divides the text of the body into educational and work experience; in Figure 8.7 we see Kathy's modification of the earlier letter to create a dynamic, concise presentation in which descriptions of the two types of experience merge into one paragraph. This can be an effective strategy if the two subjects overlap, and Kathy was very interested in demonstrating just that kind of synergy. Note the opening and closing sections of both letters and the importance of transitional elements in steering the reader toward the appropriate detail.

Exercise

Write a polished cover letter applying for the job you have targeted; keep this with the other documents in your job folder.

Figure 8.6   Four-Paragraph Cover Letter

Kathy Latham
456 Pell Court   Telephone 216-555-5555
Surrey, Illinois 63098   klatham555@aol.com

November 6, 2001

Joyce C. Armbruster
EdText, Inc.
153 Fairview Hill Road
Lynette, Illinois 60561

Dear Ms. Armbruster:

I write to apply for the job of regional sales coordinator which was posted in the November 4 *Chronicle of Higher Education*. I can supply the skills, drive, and dedication that you want.

The skills and devotion to education derive from two aspects—my work and college experience. On December 18, I will graduate from Powell College with a 4.0 average, taking a B.S. degree in management—a reflection on the eight years I was an award-winning district and store manager. But because I especially want to work with selling, developing and promoting K-8 educational materials, I concurrently took courses leading to Associate of Arts certification in elementary teacher preparation. These courses included beginning and intermediate classes in education, as well as field experiences in local schools.

Such experiences were a delight because I had been involved in local classrooms for years. As a volunteer in our Early Childhood Program, I learned about the educational needs of a diverse group of children, discovering how to use computers to challenge students and enhance their quality of life. My Teacher's Aide position was unique in that I had virtual autonomy in developing exercises, activities, and programs—and the freedom to become thoroughly acquainted with appropriate texts, kits, and software. Finally, my nomination to the vice presidency of the Surrey Park District Board gave me the chance to focus the park district's priorities on educational programs for younger children.

I welcome the opportunity to help your company grow. I can be reached at home at 555-5555 after 3 p.m. weekdays; an answering machine will take calls at other times. I can schedule an interview at your convenience with a day's notice. I look forward to hearing from you.

Sincerely,

Kathy Latham

Kathy Latham

Figure 8.7   Three-Paragraph Cover Letter

Kathy Latham
_____

                                              Telephone 216-555-5555
                                              klatham555@aol.com

456 Pell Court
Surrey, Illinois 63098
November 6, 2001

Joyce C. Armbruster
EdText, Inc.
153 Fairview Hill Road
Lynette, Illinois 60561

Dear Ms. Armbruster:

I write to apply for the job of regional sales coordinator, posted in the
November 4 _Chronicle of Higher Education_. I can supply the skills, drive, and
dedication that you want.

Such skills and devotion derive from my work and college experience. On
December 18, I will graduate from Powell College with a 4.0 average, taking
a B.S. degree in management—a tribute to the eight years I was an award-
winning retail manager. But because I especially want to work with selling,
developing and promoting K-8 educational materials, I concurrently earned
an Associate of Arts certification in elementary teacher preparation. Courses
included core classes in education as well as field experiences in local schools.
Such visits were a delight because I had been involved in local classrooms
for years. As a volunteer in our Early Childhood Program, I learned about
the educational needs of a diverse group of children, discovering how to
use computers to challenge students and enhance their quality of life. My
Teacher's Aide position was unique in that I had virtual autonomy in developing
appropriate exercises, activities, and programs—and the freedom to become
thoroughly acquainted with texts, kits, and software. Finally, my nomination
to the vice presidency of the Surrey Park District Board gave me the chance
to focus the park district's priorities on educational programs for younger children. These programs still thrive.

I welcome the opportunity to help your company grow. I can be reached at
home at 555-5555 after 3 p.m. weekdays; an answering machine will take calls
at other times. I can schedule an interview at your convenience with a day's
notice. I look forward to hearing from you.

Sincerely,

Kathy Latham

Kathy Latham

Vitae and Portfolios

In some fields, alternative forms of work-history information accompany cover letters. These may or may not replace résumés; sometimes they supplement them. The curriculum vitae, an extended document, acts as a combination of résumé, narrative of work history, and categorization of strengths. Most academic jobs require "CVs," but many grantors also want them appended to proposals applying for funding—whether or not a project is strictly academic. If you are going to become a teacher, you may also construct a vita.

In addition, such an expanded résumé can generate real benefit if you post it online. The point-and-click feature of such electronic documents allows for considerable depth to be accessible to the employer who is interested in saving money on interview costs, and who thus may be interested in more than a cursory overview of your qualifications. You can create files to be accessed by mouse-click that might include replicas of certificates, awards, and important projects mentioned in the main document. Your "CV" can become an electronic dossier, accessible at either the general level or in detail.

Figure 8.8 is a short example of such a document—variations of headings, classifications, and approach occur among these documents in different fields, but the guidebooks of professional organizations explain such nuances.

Fields relying on artistic or speaking ability may require creation of a portfolio to accompany a résumé or CV. I once had to include a portfolio in an application seeking support for a community art project. My portfolio included slides and color glossies of my work and that of my collaborators, as well as narrative summaries of our experience. Architects, photographers, and commercial artists may need to assemble portfolios displaying their best work; increasingly, teachers must submit visual portfolios in the form of cassettes recording classroom style and approach, and sales managers must supply their best presentations on video. Check with others in your profession to determine whether such supplementary materials are required during the job search. The following list offers some guidelines for preparing a portfolio.

◆ Study examples in your field.
◆ Select your documents and pictorial materials carefully.
◆ Organize the contents to fulfill expectations within your profession.
◆ Produce videos and pictures professionally.

Interviews

If the reviewers of your cover letter, résumé, and supporting material decide they want to meet you, you will be contacted and a time arranged for an interview. Because of the high costs incurred in bringing people to the site, much preliminary screening is now done by phone and e-mail. Even a rather

Figure 8.8   Vita

VITA OF: L.J. Brand
16 Barton Lane
Clinton, Virginia 15813
(301) 555-5555

WORK HISTORY
Associate Professor, Strawn College, Virginia, 1994-Present
—Vice President, Faculty Senate, 1999-2000
—Omicron Omega Honor Society Advisor, 1996-Present
—Chair, Faculty Library Committee, 1994-96
—Writing Programs Coordinator, 1994-95

Assistant Professor, Humanities, Langdon College, 1988-94
—Course development—conventional and site-based
—Planning and advisory committees
—Community writing projects
—Teaching/retention commendation by department chair

Lecturer, Fowler University, Illinois, 1984-88
—Development/teaching of literature and writing courses
—Published commendations for outstanding teaching

Teaching Assistant, University of Illinois, 1975-81

EDUCATION
Ph.D., English Literature, University of Illinois, 1981
M.A., English Literature, University of Missouri, 1975
B.A., English, University of Missouri, 1974

STRENGTHS
—Excellent teaching skills in composition and literature
—Expertise in designing traditional and non-traditional courses
—Activity in scholarship

DISSERTATION
Testing and the Quest explores Shakespeare's disruption of motifs. Director:
Harfield Burke.

HONORS
—College Alumni Teacher of the Year, 1997
—Scholarship, NIF National Institute, July 21-23, 1996
—Dean's commendation for outstanding teaching, May, 1996
—Published commendations for excellence in teaching, 1980, 1981, 1985
—NEH Research Grant, 1982

PROFESSIONAL ORGANIZATIONS
The Modern Language Association of America

Figure 8.8 *(continued)*

The Society for Literature and Science
The Mid-Atlantic Popular Culture Association

INTERESTS

Renaissance and American literature, scientific and mythological influences upon literature.

TEACHING (Sections Follow Parentheses)

Strawn College, 1994-Present

World Literature from the Renaissance (English 210)—1
Children's Literature (English 300)—2
Freshman Composition (English 101)—10
Freshman Composition (English 102)—9
Technical Writing (English 220)—6
English Literature I (English 207)—3
English Literature II (English 208)—1
American Literature (English 202)—2
American Literature (English 203)—1

Langdon College, 1988-94

Accelerated Composition (English 106)—2
Composition I (English 101)—11
Composition II (English 102)—24
British Literature I (Lit. 201)—8
British Literature II (Lit. 202)—8
Introduction to Modern Literature (Lit. 120)—7

Fowler University, 1984-88

Advanced Expository Writing (English 381)—10
Introduction to Poetry (English 101)—8
Introduction to Drama (English 102)—7
Introduction to Fiction (English 103)—12

University of Illinois, 1975-81

Freshman Composition for Engineers (Rhet. 105)—3
Freshman Composition (Rhet. 105)—10
Advanced Composition (Rhet. 143)—7

SELECTED ARTICLES AND PRESENTATIONS

"Meditation in the English Renaissance," *SRSP* 20 (1997):15-22.
"Vincentio: Problematic Origins" in *Proceedings: Problem Plays*. Ed. David Deane. Foster: Foster UP, 1997.
"Chaos Theory and the Dark Comedies," paper for the Virginia Shakespeare and Renaissance Conference, Charlottesville, April 10-12 1997.
"Science-Fiction and Shakespeare," presented at the "Literature *Plus* Science" Conference, Atlanta, GA, October 10-13, 1996.

subdued call from a personnel office asking whether you can come for a visit might be an interview in itself; be at your best and be sure to take notes about important issues. In industry, two personal interviews are often conducted— a preliminary one (at a college or regional center, perhaps) and then a site visit. Perhaps there will be a return visit. It's difficult to keep calm and organized during the interview process, but if you've done your homework, you will ask the right questions and provide the right answers. Here are some points to remember about this stage in the job-hunting process:

1. *Know* as much as possible about the company.
2. *Make a list* of questions that you are probably going to be asked. Many interviewers ask some form of the questions presented in Figure 8.9.
3. *Practice* ahead of time—with friends asking the likely questions.
4. *Be well-rested* before the interview.
5. *Stay calm* but attentive during the interview; don't fidget and don't interrupt the interviewers.
6. After the interviewers have asked you questions, they will frequently inquire whether you have questions of them. Be sure to *ask about that which you need to know.*
7. At the close of the interview, *leave the impression of availability and desire to work there*—unless you have decided that that company is not the place for you, as sometimes happens.

Figure 8.9 Frequently Asked Interview Questions from the Author's Own Interview Notes

1. Why have you applied to work for us?
2. Explain how your work experience qualifies you for this job.
3. Explain how your educational background qualifies you.
4. Envision a typical day here; describe the things you think you will do.
5. Five and ten years from now—what do you believe you will be doing?
6. What plans do you have to keep current in this field?
7. What questions do you have about us?
8. Describe a problem you encountered in your previous job, and explain how you solved it.
9. What decision process has led you to this career?
10. How might you benefit this firm?
11. What did you like best in (college or) your last job? What least? Why?
12. Define satisfaction. When are you satisfied with what you have been doing?
13. What are you looking for in a company?
14. Will this change?
15. How would you like to manage/be managed?

Exercise

In two-to-four-person small groups in class, trade information about the jobs you have targeted for research. Take turns interviewing one another using the lists of questions you have generated, prompted by the questions shown in Figure 8.9. Discuss the strengths and weaknesses of the responses *and* the questions.

Follow-up Correspondence

After returning from an interview, type a short letter that thanks the interviewers, expresses interest in a specific topic discussed during the interview (something that highlights a skill you have), and leaves the reader feeling that you are very interested in the job (unless that is no longer the case). A formally-written e-mail is useful in addition to the follow-up letter, and may, when time is of the essence, suffice by itself—but you should adhere to the practice of your specific field in determining that.

Figure 8.10 depicts a typical follow-up letter. Kathy *wanted* this job!

Figure 8.10   Follow-up Letter

Kathy Latham

456 Pell Court   Telephone 216-555-5555
Surrey, Illinois 63098   klatham555@aol.com
November 18, 2001

Joyce C. Armbruster
EdText, Inc.
153 Fairview Hill Road
Lynette, Illinois 60561

Dear Ms. Armbruster:

Thank you for your interview with me yesterday. I enjoyed our discussions and know I would have a stimulating part to play in your company's evolution. We certainly have come a long way from the days of boxed reader kits!

I was excited to learn more about how software can be custom-created for language programs and to realize that my skills can enhance such a project. One of my college projects was, in fact, the creation of an instructional CD-ROM; another was the writing of an animated Web page.

I deeply appreciate your hospitality. Please let me know when you make your decision.

Sincerely,

Kathy Latham

Kathy Latham

Exercise

Write a follow-up letter responding to your in-class interview.

Assignments

1. Review and polish all of the material you have written and accumulated as you have studied and targeted a hypothetical job. Organize it into a portfolio for your future reference. Include the job posting itself; the notes taken in researching this job; a cover letter; your best résumé, perhaps with a separate list of references; a list of potential questions for an interview; and a follow-up letter. Your instructor may request this portfolio as a graded assignment.

2. Take this assignment to the next step—search the Internet for a hypothetical job opening in your field—in another country. Consider strategies covered in this text concerning international communications, and create a job application portfolio consisting of

   a. A preliminary e-mail response to the advertisement, requesting appropriate materials and a more detailed description of the position

   b. A fax message with the following components:
      * A fax cover sheet, filled out clearly and containing a short summary of the material to follow
      * A formal letter of application using "world English" clearly and concisely (See Figure 8.11 for tips on international communication).
      * A résumé that the fax machine can "read" clearly
      * A list of references

Figure 8.11 Checklist for International Communications

❑ Am I writing or speaking in "world English"?

❑ Am I using key words in the field as reference points?

❑ Am I avoiding jargon?

❑ Is my word order predictable?

❑ Does text come with an alternative translation?

❑ Am I using graphical aids to clarify meaning?

❑ Am I providing ample time and space for responses?

❑ When presenting orally, will I also provide a written text?

c. An e-mail message in "world English" responding to a request for further information, namely a short sample of a research project you have done, a marketing proposal you have produced, or another formal report on something in your field (You might use or modify your multi-sectioned technical report for this). The e-mail should explain that you will fax the material to the recipient at a certain time, and should thank the recipient for his or her continued interest in you as a job candidate.

d. A fax cover sheet plus the formal report or project

e. An e-mail response to a communication from the company requesting that you schedule an interview with the Chicago branch of its subsidiary

3. Now, take this quest to the next level. Imagine you are to prepare a presentation to members of that company's marketing, engineering, in-house training, or public-relations team. Classmates in your small groups will take on the roles of that team: a group of ten makes an ideal size and can be formed by combining two small groups. Using the strategies and techniques discussed in the text, create a dynamic presentation that considers timing, enunciation, and clarity of delivery. Practice this presentation in front of a mirror, taping yourself to ensure that your delivery works. Plan the presentation using

a. Audience analysis (Chapters 1 and 2)

b. Storyboards, task lists, and other invention strategies (Chapter 5)

c. A researched concept with supporting detail (Chapter 7)

d. Graphics for visual impact (Appendix)

Then deliver the presentation to your group.

4. Figure 8.12 shows the text of an electronic résumé that contains links to other supporting documents a recruiter might need to see. The résumé tags important words so that clicking on them will send the reader to supporting files of interest. Assume that you can click on those words in this figure—what features do you want in the supporting files? How might each file be organized and presented, and which files can be linked to the same words?

5. If technical assistance permits, construct an online presentation including a résumé and a link to your own major project, or modify the process of Questions 1–3 by using e-mail attachments instead of faxes. Send the e-mail messages to your classmates who will be your "interviewers."

Figure 8.12   Text of Electronic Résumé

RÉSUMÉ OF

Martin G. Ramirez
9 Wedgewood Circle
Freedonia, MO 63136
(314) 444-5555
mramirez444@hotmail.com

~Please click on active files below~

OBJECTIVE—Managerial position in a metals manufacturing company utilizing my credentials in engineering, mathematics, and business management.

SKILLS—Product development and supervision
Metallurgical engineering
Division management
Organization

EDUCATION—M.B.A., Northern Illinois University
Ph.D., Metallurgical Engineering, University of Illinois
M.S., Metallurgical Engineering, University of Illinois
B.S., Mathematics, University of Utah

WORK EXPERIENCE—Product Development Supervisor, Missouri Metals
Development Engineer, Rolled Products, Missouri Metals

PUBLICATIONS—Please click here for this file.

REFERENCES—Please access through e-mail address above.

References

Introduction

There are many sources to help you investigate the principles and practice of technical writing. I've listed just some below, and have added annotations to describe them.

American Psychological Association. *Publication Manual of the American Psychological Association*. 5th ed. Washington, DC: APA, 2001. Explains APA style and provides models for its use. One may also access the APA Web site (www.apastyle.org) for helpful material.

Borowick, Jerome N. *Technical Communication and Its Applications*. 2nd ed. Englewood Cliffs, NJ: Prentice-Hall, 2000. A guide for professionals and students, using and critiquing engineering, scientific, and business documents.

Brereton, John C., and Margaret A. Mansfield. *Writing On the Job: A Norton Pocket Guide*. New York: Norton, 1999. Discusses basic business and technical documents; also includes a section on journalistic writing.

Brusaw, Charles T., Gerald J. Alred, and Walter E. Oliu. *Handbook of Technical Writing*. 7th ed. New York: St. Martin's, 2003. A large reference manual arranged as a dictionary of terms.

Carbone, Nick. *Writing Online: A Student's Guide to the Internet and World Wide Web*. 3rd. ed. Boston: Houghton Mifflin, 2000. A how-to and resource guide for working and writing in cyberspace.

Gibaldi, Joseph. *MLA Handbook for Writers of Research Papers*. 6th ed. New York: Modern Language Association, 2003. Explains MLA style and provides models for its use.

Harris, Muriel. *Prentice Hall Reference Guide to Grammar and Usage*. 5th ed. Upper Saddle River, NJ: Prentice Hall, 2003. An excellent guide to style, mechanics, and format.

Holloway, Brian R. *Proposal Writing Across the Disciplines*. Upper Saddle River, NJ: Prentice Hall, 2003. This book contains further information about writing proposals that respond to a variety of business and academic needs.

Irvine, Martin. *Web Works.* Norton Pocket Guides. New York: Norton, 1997. A hard-working compact guide to using the Internet. Contains a helpful glossary.

Krull, Robert, ed. *Word Processing for Technical Writers.* Baywood's Technical Communications Series. Vol. 3. Amityville, NY: Baywood, 1988. Though dated regarding technical specifics, this book does a good job of alerting the reader to editorial and format problems that occur when writing with machines.

Lannon, John M. *Technical Communication.* 9th ed. New York: Longman, 2003. A big text strong in its use of graphics.

Locke, Lawrence F., Waneen Spirduso, and Stephen Silverman. *Proposals That Work: A Guide for Planning Dissertations and Grant Proposals.* 4th ed. Thousand Oaks, California: Sage, 2000. A useful guide for the graduate student.

Pearsall, Thomas E. *The Elements of Technical Writing.* 2nd ed. The Elements of Composition Series. Boston: Longman, 2001. A short manual covering writing basics, format, and ethics.

Pfeiffer, William S. *Technical Writing: A Practical Approach.* 5th ed. Columbus, OH: Prentice Hall, 2003. A thorough text focusing on writing in a corporate context.

Reep, Diana C. *Technical Writing: Principles, Strategies, and Readings.* 5th ed. Boston: Allyn and Bacon, 2003. A large text reinforced by a grammar guide.

Turabian, Kate L., John Grossman, and Alice Bennett. *A Manual for Writers of Term Papers, Theses, and Dissertations.* 6th ed. Chicago: U of Chicago P, 1996. A revised edition of a helpful handbook based on University of Chicago style that contains much valuable advice for the writer.

Ward, Dean. *Tradition and Adaptations: Writing in the Disciplines.* Columbus, OH: McGraw Hill, 1997. Not a book devoted to technical writing *per se*, but instead a text covering advanced documents in the professions.

Woolston, Donald C., Patricia A. Robinson, and Gisela Kutzbach. *Effective Writing Strategies for Engineers and Scientists.* Chelsea, MI: Lewis, 1988. An older book, but thought-provoking in its discussion of the legal implications of scientific and engineering writing.

Valediction

As you can see by perusing this bibliography and as a result of your work in this course, the subject of business and technical writing—or transactional writing, as it is perhaps best called—is enormous. In this text, you've focused on certain key aspects within the large totality. You have reviewed basic patterns of construction and mechanics, then put these to work—first drafting letters and memos, then expanding your tasks to include announcements, process descriptions, short reports, and formal reports. Finally, you have experimented with creating the job-finding materials (cover letters, résumés, and related documents) that you will be writing at different points in your working life. May that life be a long, creative one, enhanced by your transactional writing skills!

Enhancing Your Document with Graphics

Purpose of Visual Displays

A picture may not really be worth a thousand words, but it can make or break a presentation nonetheless. Some types of written communication—and oral presentation—almost *demand* charts, graphs, or other pictorial material. This has been true since ancient times; note the example of process description illustrating a technique in butchering (Figure A.1). It came from Old Kingdom Egypt, approximately 2500 BCE.

What was true in the ancient world is true today, and more critical—since our attention spans have been conditioned by television, computer multimedia, and a remarkably fast pace of living. A list of statistics may not impress since it takes too long to interpret; the same statistics displayed on a graph or in a table reveal instantly the internal logic, structure, or principle at work.

Tables

Tables provide a means of depicting such logical relationships. Note Figure A.2, a table explaining to carpenters the correct opening size to cut in order to install an attic ladder. Compare that clear display of information with this hypothetical prose equivalent: "For model numbers W-2208, W-2210, WH-2208, and WH-2210, cut a rough opening 22^1/$_2$ inches by 54 inches; for models W-2508, W-2510, WH-2508, and WH-2510, use a 25-inch by 54-inch rough opening."

Obviously, the table helps the reader interpret the data. Such tables may be charts of descriptive phrases as well as grids containing numbers. For better depiction of the internal logic of data, however, one may need graphs instead.

Graphs

Graphs normally depict two variables interacting. Because the graphs that you will most likely use show two things plotted against each other—stress as a function of time, stock prices as a function of inflation—they are essentially *two-dimensional*. Bar graphs may show the rise and fall of something discrete

Figure A.1   An Ancient Illustration of Process

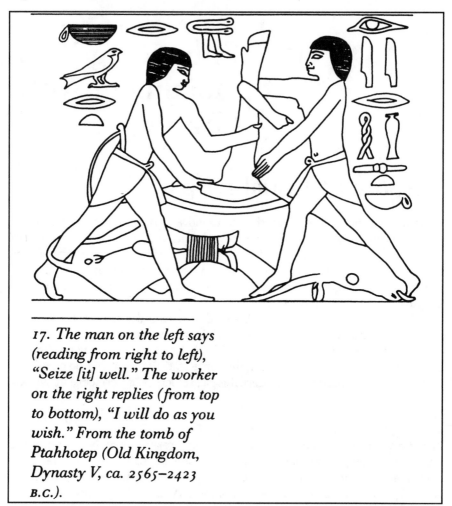

17. The man on the left says (reading from right to left), "Seize [it] well." The worker on the right replies (from top to bottom), "I will do as you wish." From the tomb of Ptahhotep (Old Kingdom, Dynasty V, ca. 2565–2423 B.C.).

Source: Silverman, David. *Language and Writing in Ancient Egypt*. Pittsburgh: The Carnegie Museum of Natural History, 1990, page 31.

over time to clarify a comparison; line graphs may plot a continuous change of something over time. In both cases, time might be plotted on the horizontal axis, and the "something that changes" might be plotted on the vertical axis. Figures A.3 through A.5 exemplify these kinds of graphs. Contrast helps the comparative aspect of bar graphs, as Figure A.3 demonstrates—consider if the two items compared were to appear in the same shade of gray. Although a simple monochrome line graph such as that of Figure A.4 can be effective, adding color filler below the line will make your illustration more memorable, as is suggested by Figure A.5.

Figure A.2 A Table Clarifying the Text

(Continued)

CUTTING A HOLE IN THE CEILING

STEP 5. Draw a rectangle the size of the rough opening on the ceiling, with one edge parallel to a joist (See Figure 7). You may do this by sawing until you reach a joist and using it as a frame of reference. (The size of the rough opening is shown on the box or in Table 2).

Note: Locating at least one edge of the opening along a ceiling joist will allow the joist to be used as a side of the frame you will build. This will simplify framing the rough opening.

STEP 6. Cut out the rest of the ceiling within the marked outline following these instructions:

a. **Do not cut any joists at this time.** Cut through the ceiling only.

b. Remove the ceiling in small pieces because ceiling material can be very heavy.

STEP 7. If no joists span the hole in the ceiling, go to Section 7 "FRAMING THE ROUGH OPENING" on page 12.

If any joists span the hole, go to Section 6 "CUTTING THE CEILING JOISTS" on the next page.

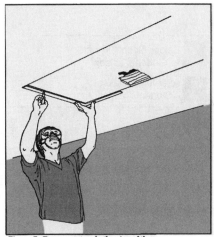

Figure 7. Draw a rectangle the size of the rough opening.

Model Number	Rough Opening Size
W 2208 W 2210 WH 2208 WH 2210	22-1/2" × 54"
W 2508 W 2510 WH 2508 WH 2510	25" × 54"

Table 2. Rough Opening Size

Another visual aid related to the line or bar graph is the *pie graph* (Figure A.6), used to depict such things as the different portions of a budget, the percentage of teenagers who are juvenile delinquents, or the amount of total telephone charges taxed. Here, one variable is replaced by a constant, which is the whole item, or 100 percent; the graph shows what divisions of 100 percent are allocated to the different components. Such illustrations stand midway between graphs and tables, and are only efficient in showing components of a whole. To depict changes over time might require several pies, or a whole bakery!

Figure A.3   Bar Graphs Can Relate Two Concepts over Time

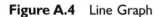

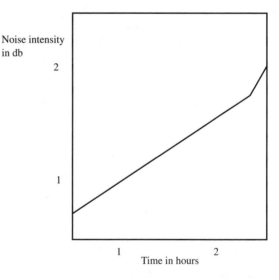

Figure A.4   Line Graph

Drawings

Illustrations that contain more pictorial content than tables and graphs present different challenges to the technical communicator and provide different options depending upon whether the communication is a written report or an oral presentation.

Figure A.5   Filler in Line Graph

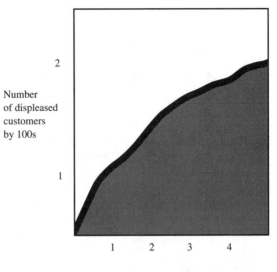

Number
of displeased
customers
by 100s

Noise duration in hours

Figure A.6   Pie Graph with Contrast Used to Emphasize Differences

Daily Routines of Observed Subjects

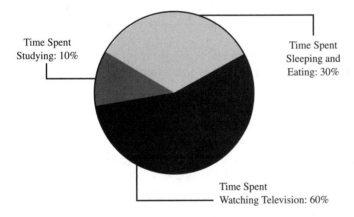

Time Spent
Studying: 10%

Time Spent
Sleeping and
Eating: 30%

Time Spent
Watching Television: 60%

Options for Writing: The writer might include freehand illustrations in the text of a report, or freehand illustrations enhanced by the use of engineering templates. Such plastic templates contain circuit diagrams, ovals, circles, rectangles, and a variety of other contours including those of Erlenmeyer flasks. French curves—plastic curves with continually varying radii—are handy as well. In the past, the well-equipped draftsperson might own a sizable collection of such plastic shapes. Even now, they are useful, although computer programs have taken over many of the duties formerly assigned to the odd-looking acetate stencils.

Another option for the technical writer is photocopying an existing illustration, enlarging or reducing that picture as needed, then cutting away the excess border and gluing or taping the illustration in its proper place in the report. Either a "hole" can be left in the text for this purpose, or a location on a blank sheet can be printed upon. The finished "paste-up" may be photocopied again for a clear, crisp look. Rubber cement, a glue stick, or a high grade of clear tape work well for this purpose, although many brands of tape and cement discolor over time. Remember to cite any visual material that is not yours and to request permission for its use if employing it for commercial purposes.

A modern variation on this tedious procedure is to use a flatbed scanner to scan the illustration into your document. The scanner's software should allow you to manipulate the size and appearance of the image. Again, remember that any visual material that is not yours needs to be credited in a caption and on any reference sheet you supply; commercial use will require written permission from the source or copyright holder.

Computer Art: The illustrator's options now include pre-made art resident in a file in the computer, clip art and photographs accessible online, and additional art purchased in disk form from software vendors. Much of this art can be modified by bringing it up in a paint or draw program. Some of this art is in the public domain and therefore does not require citation. Remember, though, that proprietary material not in the public domain is the property of its original creator, whether or not such art is altered, and therefore requires citation. Permission must be granted for the commercial use of proprietary art as well. Figure A.7 gives options in using artwork in a report.

In addition, drawing tools contained in many programs allow you to create your own art and import it into the document you construct. (Such

Figure A.7   Some Handy Art Options

Long written reports benefit from the use of artwork to reinforce a point. Remember that several techniques can introduce visual material into a text:

- Reproduce the figure on a high-resolution copier, enlarging it or reducing it as needed; then, carefully cut the copied figure to size and glue it with rubber cement to the appropriate spot on the text.
- Copy the figure on a high-resolution machine after first positioning it where it must appear on the page. Then arrange for your text to have a "hole" into which the visual will fit. When printing that page of the report, place the sheet with the figure on it into the printer hopper and the text will print around the figure. Make several practice spares first!
- Call up the appropriate figure from the clip art or drawing part of your word-processing program, then insert per the program's instructions.
- Scan the figure into your document using a flatbed scanner and an optical-recognition program.

drawings do not need to be overly linear and sticklike, either; sketching pro-grams can produce charcoal, chiaroscuro, and other sophisticated effects).

Programs such as those by Adobe and others allow the illustrator to call up digitally stored photographs and manipulate them, editing for best effect in the document. Again, if you use a commercial disk of photographs, you must credit the source and acquire permission granting for-profit use.

Options for Presentations

The presenter certainly can exploit much of what is available to writers, but has a number of other helpful visual options to assist oral delivery.

Charts and Displays

In addition to the well-worked and overused chalkboard or the more modern dry-erase board, consider these other tools:

1. *Portable flip-pads* with poster-size paper can be handy—especially to record questions for response. Print in large letters and use dark markers in deference to members of your audience sitting in the back of the room (Figure A.8).

Figure A.8   A Flip Chart Can Be a Basic Visual Enhancement

Figure A.9   Large Posters Can Be Tiled Across the Presentation Area

2. *Posters* themselves, arranged in sequence or attached in a large flip-chart, can focus visual attention. Computer programs can generate these posters in varying sizes and can create a sequence of large, attractive visual "tiles." These may work as a backdrop for the speaker—one that the speaker may point to and reference—or they may also provide an exhibit at a booth. The sequence may be used as a free-standing storyboard (Figure A.9).

Mechanical Devices

Use these as needed; remember that the presentation must dominate, rather than the number or frequency of use of mechanical aids. Avoid, for example, the basic and bulky *opaque projector* which, though accommodating books, pages, and sheets, has the disadvantage of requiring frequent refocusing and manipulation of the material. The following options yield more favorable results:

◆ An old standby in business and teaching, the *overhead projector* requires transparencies rather than opaque sheets—thereby obligating one to prepare such transparencies carefully. If making a transparency from an illustration surrounded by text, do not do so directly, unless you want the text; rather, make a single paper copy of the material using a high-resolution photocopier. Next, trim the excess material from the illustration, and produce the transparency from this "clean" copy. This way, nothing will deflect the audience's attention from the visual. Color transparencies can be made on color copiers: the best of these machines produce deep, saturated colors that project well. Special transparency material can be run through your computer printer as well; use only blanks compatible with your machine.

◆ *Slide projectors* can be activated by remote switches, freeing the presenter from hovering near an overhead device. Make sure that you check the orientation of your slides ahead of time so that none are backward or upside-down. Effective slides can be made using a computer program to create a series of stills. Slides can be produced from drawings rendered on the computer, clip art, or scanned-in items.

◆ *Computer projection* programs contain animating functions as well, enabling you to create presentations in which items fade, dissolve, or sequence to the next display. This computer-generated material may be used with a projector in conjunction with a laptop device. These machines require darkness and good connections between all parts of the equipment; always have something to do in case of a "glitch" (e. g., come prepared with handouts featuring copies of the material). Be sure not to let the visual tricks overshadow the point of the presentation. Repeated use—or distribution—of such material probably mandates its storage on a read/write CD-ROM.

◆ Such animated material may be shown on a *television* equipped with a special adapter connected to your computer. Conventional videos can be used effectively in presenting, as well; if using videotape, make sure it is well-edited and cued to begin at the appropriate place.

Index

213